DISASTERS by Design

A Reassessment of Natural Hazards in the United States

DENNIS S. MILETI
University of Colorado at Boulder

with the contributions of participants in the Assessment of Research and Applications on Natural Hazards

AN ACTIVITY OF THE
INTERNATIONAL DECADE FOR
NATURAL DISASTER REDUCTION

Joseph Henry Press
Washington, D.C.

JOSEPH HENRY PRESS • 2101 Constitution Avenue, N.W. • Washington, D.C. 20418

The Joseph Henry Press, an imprint of the National Academy Press, was created with the goal of making books on science, technology, and health more widely available to professionals and the public. Joseph Henry was one of the founders of the National Academy of Sciences and a leader of early American science.

Library of Congress Cataloging-in-Publication Data

Mileti, Dennis S.
Disasters by design : a reassessment of natural hazards in the United States / Dennis S. Mileti
p. cm. — (Natural hazards and disasters)
Includes bibliographical references and index.
ISBN 0-309-06360-4 (alk. paper)
1. Emergency management—United States. 2. Disaster relief—United States. 3. Natural disasters—United States. 4. Hazardous geographic environments—United States. 5. Sustainable development—United States. I. Title. II. Series.
HV551.3 .M55 1999
363.34'0973—ddc21 99-29511
CIP

Printed in the United States of America.

Contributing Authors

Christopher R. Adams, Colorado State University
Daniel Alesch, University of Wisconsin at Green Bay
James Ament, State Farm Fire and Casualty Co.
Jill Andrews, University of Southern California
Norbert S. Baer, New York University
Jay Baker, Florida State University
Conrad Battreal, U.S. Army Corps of Engineers
Timothy Beatley, University of Virginia at Charlottesville
Stephen O. Bender, Organization of American States
Philip R. Berke, University of North Carolina at Chapel Hill
Dennis Black, Colorado State University
B. Wayne Blanchard, Federal Emergency Management Agency
Robert Bolin, Arizona State University
Patricia A. Bolton, Battelle Human Affairs Research Center
Linda B. Bourque, University of California at Los Angeles
David Brower, University of North Carolina at Chapel Hill
William M. Brown, U.S. Geological Survey
Ian G. Buckle, State University of New York at Buffalo
Angelia P. Bukley, National Aeronautics and Space Administration
Raymond J. Burby, University of New Orleans
Bonnie Butler, Federal Emergency Management Agency
David L. Butler, University of Colorado at Boulder
Peggy A. Case, Terrebonne Readiness and Assistance Coalition

Harold Cochrane, Colorado State University
Nicolas Colmenares, University of Colorado at Boulder
Louise Comfort, University of Pittsburgh
James N. Corbridge, Jr., University of Colorado at Boulder
Susan L. Cutter, University of South Carolina at Columbia
JoAnne DeRouen Darlington, Colorado State University
James F. Davis, California Division of Mines and Geology
Robert E. Deyle, Florida State University
John A. Dracup, University of California at Los Angeles
Thomas Durham, Central U.S. Earthquake Consortium
Ute Dymon, Kent State University
Charles Eadie, City of Watsonville at California
David Etkin, Environment Canada
Dean C. Flesner, State Farm Fire and Casualty Co.
Betsy Forrest, University of Colorado at Boulder
Alice Fothergill, University of Colorado at Boulder
Steven P. French, Georgia Institute of Technology
Karen Gahagan, Institute for Business and Home Safety
Luis A. Garcia, Colorado State University
David F. Gillespie, Washington University
David R. Godschalk, University of North Carolina at Chapel Hill
Joseph H. Golden, National Oceanic and Atmospheric Administration
Marjorie Greene, Earthquake Engineering Research Institute
Eve Gruntfest, University of Colorado at Colorado Springs
Neil Hawkins, University of Illinois at Urbana-Champaign
Walter Hays, U.S. Geological Survey
James P. Heaney, University of Colorado at Boulder
Edward J. Hecker, U.S. Army Corps of Engineers
Henry Hengeveld, Environment Canada
Michael Hodgson, University of South Carolina at Columbia
Ken Hon, U.S. Geological Survey
Dale Jamieson, University of Colorado at Boulder
Brian Jarvinen, National Hurricane Center
Edward Kaiser, University of North Carolina at Chapel Hill
Jack D. Kartez, University of Southern Maine
Robert Klein, National Association of Insurance Commissioners
Howard Kunreuther, University of Pennsylvania at Philadephia
James F. Lander, University of Colorado at Boulder
Larry Larson, Association of State Floodplain Managers
Eugene L. Lecomte, Institute for Business and Home Safety

Michael K. Lindell, Texas A&M University
Rocky Lopes, American Red Cross
George Mader, Spangle Associates
Enrique Maestas, University of Texas at Austin
Robert Martin, University of California at Berkeley
William Martin, ITT Hartford
Peter J. May, University of Washington at Seattle
James McDonald, Texas Tech University
Elaine McReynolds, Federal Emergency Management Agency
Kishor Mehta, Texas Tech University
Mario Mejia-Navarro, Colorado State University
Jerry Mitchell, University of South Carolina at Columbia
Elliott Mittler, University of Wisconsin at Green Bay
Jacquelyn Monday, JLM Associates
Betty Morrow, Florida International University
John Mulady, United Services Automobile Association
John-Paul Mulilis, Pennsylvania State University at Monaca
Susan Murty, University of Iowa at Iowa City
Mary Fran Myers, University of Colorado at Boulder
Sarah K. Nathe, California Office of Emergency Services
David Neal, University of North Texas
Joanne M. Nigg, University of Delaware
Stuart Nishenko, Federal Emergency Management Agency
Franklin W. Nutter, Reinsurance Association of America
Eric K. Noji, Centers for Disease Control and Prevention
Charles Nyce, University of Hartford
Paul W. O'Brien, California State University at Stanislaus
Robert Olshansky, University of Illinois at Urbana-Champaign
Robert Olson, Robert Olson Associates, Inc.
Philip N. Omi, Colorado State University
Michael J. O'Rourke, Rensselaer Polytechnic Institute
Risa Palm, University of North Carolina at Chapel Hill
Eve Passerini, University of Colorado at Boulder
Edward Pasterick, Federal Emergency Management Agency
Pamela Pate, University of Texas at Austin
Robert Paterson, University of Texas at Austin
Walter Gillis Peacock, Florida International University
Ronald Perry, Arizona State University
William J. Petak, University of Southern California
Jon A. Peterka, Colorado State University

Brenda Phillips, Texas Woman's University
John C. Pine, Louisiana State University
Rutherford H. Platt, University of Massachusetts at Amherst
Roy S. Popkin, Popkin Associates
Kenneth W. Potter, University of Wisconsin at Madison
George Rogers, Texas A&M University
Richard J. Roth, Jr., California Department of Insurance
Richard J. Roth, Sr., Consultant
Claire B. Rubin, Claire B. Rubin & Associates
David Sample, University of Colorado at Boulder
Paula Schulz, California Office of Emergency Services
Stanley Schumm, Colorado State University
James Schwab, American Planning Association
Michael S. Scott, University of South Carolina at Columbia
William Solecki, Florida State University
John H. Sorensen, Oak Ridge National Laboratory
Kenneth R. Stroech, U.S. Environmental Protection Agency
Richard Stuart, U.S. Army Corps of Engineers
Craig E. Taylor, National Hazards Management, Inc.
Deborah S. K. Thomas, University of South Carolina at Columbia
John Tiefenbacher, Southwest Texas State University
Kathleen Tierney, University of Delaware
L. Thomas Tobin, Tobin and Associates
Kenneth C. Topping, Cambria Community Services Department
Susan K. Tubbesing, Earthquake Engineering Research Institute
Eric VanMarcke, Princeton University
Barbara Vogt, Oak Ridge National Laboratory
Dennis Wenger, Texas A&M University
James L. Wescoat, University of Colorado at Boulder
French Wetmore, French & Associates, Ltd.
David Whitney, California State University at Long Beach
Berry Williams, Berry A. Williams & Associates
Leonard T. Wright, University of Colorado at Boulder

Foreword

THE NATION'S FIRST ASSESSMENT of research on natural hazards got under way in 1972 at the Institute of Behavioral Sciences of the University of Colorado at Boulder. It was an interdisciplinary effort, with an emphasis on the social sciences, funded by the National Science Foundation and led by geographer Gilbert F. White and sociologist J. Eugene Haas. The project involved many graduate students, scholars, practitioners, and policymakers. Their aim was to assess the nation's knowledge of natural hazards and disasters, with an emphasis on the social sciences. Along the way they expected to point out major needed policy directions for the nation and to inventory future research needs. A summary volume issued at the completion of the project (White and Haas, 1975) brought the substance of that knowledge together, analyzed the gaps, and recommended certain policy changes and numerous research needs. (A retrospective look at the impacts of that assessment is presented in Appendix B.)

Contemporary conversations about how sustainable hazards mitigation could result in disaster-resilient communities began in the early 1990s among a few individuals who worked in federal agencies and academia. It was formalized during a workshop in the summer of 1992 in Estes Park, Colorado. Attended by over five dozen of the nation's leading hazards experts, the workshop concluded that it was appropriate to move forward with a second assessment of hazards in the United States and that the unifying theme for the work should be sustainable development.

A subsequent workshop in Boulder, Colorado, in October 1994 brought many of the same people and others together to create a specific agenda for the second assessment. At that time, members of the nation's policymaking arena—for example, the Subcommittees on Natural Hazards and Risk Analysis in the White House and the Federal Emergency Management Agency—explored the idea of further linking hazards and sustainability. These discussions and the work that followed from them have led to the present volume.

The many experts who contributed to this project started by going to work on the project's formal mission. This was to summarize what is known in the various fields of science and engineering that is applicable to natural and related technological hazards, and to make some research and policy recommendations for the future. The project began with its feet firmly planted in science and engineering, but it ended in an ontological expression. The single most important contribution that this second assessment has to offer is the recommendation for a fundamental shift in the character of how the nation's citizens, communities, governments, and businesses conduct themselves in relation to the natural environments they occupy. This book calls for and speaks to the specifics required to shift the national culture in ways that would stop at its genesis the ever-increasing spiral of losses from natural and technological hazards and disasters. The task will be to create and install "sustainable hazards mitigation" in the culture of the nation.

Disasters by Design does not stop at summarizing the hazards research findings from the past two decades. Instead, it synthesizes what is known and proceeds from that synthesis to outline a proposed shift in direction in research and policy for natural and related technological hazards in the United States.

This book takes a broad focus and occasionally a somewhat speculative tone. It is aimed at a general audience, including policymakers and practitioners. Scholarly readers will note that there are relatively few citations to academic work. This is intentional: first, because the logistics of attempting to cite everything published in the past 20 years would be overwhelming and would increase the manuscript's length by well over

100 pages. The second reason is that this volume comprises synthesized statements of what is known, collectively, about hazards and human coping strategies, which in most cases cannot properly be attributed to a specific source. As a rough way of pointing interested readers toward more information, some citations to the published literature are included at key points in the text. These should not be taken as the authoritative sources on the topics in question but simply as a starting point for more intensive reading. Readers who are interested in viewing the full list of references brought together to support this assessment can find the bibliography on the World Wide Web home page of the Natural Hazards Research and Applications Information Center at http://www.colorado.edu/hazards.

This volume is one in a series published by the Joseph Henry Press that stemmed from the second national assessment on natural and related technological hazards and disasters. The participants in the second assessment's subgroups were invited to write their own books on specialized topics and to have those manuscripts considered for publication as part of the series. The result of those invitations is several additional books in the Joseph Henry Press series. The other volumes cover insurance, organized by Howard Kunreuther and Richard J. Roth, Sr. of the Wharton School; land use, assembled by a team headed by Raymond Burby of the University of New Orleans; disaster preparedness, response, and recovery, assembled by Kathleen Tierney of the Disaster Research Center at the University of Delaware; and a national hazards and risk assessment, led by Susan Cutter at the University of South Carolina. Many other written products have and will emerge from the second assessment, but the Joseph Henry Press book series represents the second assessment's major written products.

The second assessment also resulted in the first new publication series by the Natural Hazards Research and Applications Information Center at the University of Colorado at Boulder, since the center was founded in 1976. The new series is known as the "Natural Hazards Informer." It is designed to summarize in plain language what is known in focused areas of knowledge regarding natural and related technological hazards and disasters. The series is designed to reach local practitioners, and its publication will begin in early 1999. It will be published several times a year by the center in a short newsletter-type format and will be disseminated free to some 15,000 subscribers to the center's newsletter, the *Natural Hazards Observer*.

Dennis S. Mileti
Boulder, Colorado

Acknowledgments

THIS PROJECT WAS FUNDED by the National Science Foundation under grant number CMS93-12647 with supporting contributions from the Federal Emergency Management Agency, the U.S. Environmental Protection Agency, the U.S. Forest Service, and the U.S. Geological Survey. The support of these agencies is greatly appreciated. Only the author, however, is responsible for the information, analyses, and recommendations contained in this book. A very special "thank you" is extended to J. Eleonora Sabadell and William A. Anderson of the National Science Foundation for placing their confidence in us to carry out this mission.

Special appreciation is also extended to the persons and the organizations they represent for the time and ideas they contributed as members of the project's advisory panel. Deepest gratitude is expressed to the following persons:

William A. Anderson, National Science Foundation;

Michael Armstrong, Federal Emergency Management Agency;

Riley Chung, National Institute of Standards and Technology;
Caroline Clarke, National Research Council;
James F. Davis, California Division of Mines and Geology;
Walter Hays, U.S. Geological Survey;
Edward J. Hecker, U.S. Army Corps of Engineers;
William H. Hooke, National Oceanic and Atmospheric Administration;
Robert Kistner, Colorado Office of Emergency Management;
Richard Krimm, Federal Emergency Management Agency;
Eugene L. Lecomte, Institute for Business and Home Safety;
James Makris, U.S. Environmental Protection Agency;
J. Eleanora Sabadell, National Science Foundation;
James M. Saveland, U.S. Forest Service;
Kenneth R. Stroech, U.S. Environmental Protection Agency;
Randall Updike, U.S. Geological Survey;
Gilbert F. White, University of Colorado at Boulder; and
Arthur Zeizel, Federal Emergency Management Agency.

Some of the work performed for this assessment was accomplished by subgroups of people who worked on specific topics. Special appreciation is extended to those who coordinated the work of the subgroups.

David F. Gillespie, Subgroup on the Interactive Structure of Risk;
Raymond J. Burby, Subgroup on Land Use;
James P. Heaney and Jon A. Perterka, Subgroup on Engineering;
Susan Cutter, Subgroup on Risks, Losses, and Costs;
John H. Sorensen, Subgroup on Prediction, Forecast, and Warning;
Kathleen Tierney, Subgroup on Preparedness and Response;
Dennis Wenger, Subgroup on Reconstruction;
Howard Kunreuther, Subgroup on Insurance; and
Michael K. Lindell, Subgroup on Adoption and Implementation.

Many professionals were invited in the latter part of 1997 to review the first full draft of the manuscript. The warmest regards and greatest appreciation are extended to the people listed below, who returned their useful and insightful critiques and recommendations for revision of the original draft of the manuscript:

Benigno Aguirre, William A. Anderson, Ken Baechel, Philip R. Berke, Robert Bolin, Pete Brewster, Neil R. Britton, James Bruce, Louise Comfort, Susan L. Cutter, James C. Douglas, Thomas E. Drabek, Russell R. Dynes, David Etkin, Robert Fletcher, Betsy C. Forrest, David F. Gillespie, Joseph Golden, Ruby I. Harpst, Janet C. Herrin, Kenneth Hewitt, Charles W. Howe, Wilfred D. Iwan, Paul Jennings, Steven J. Jensen, Joseph G. Kimble, Gary A. Kreps, Brett Kriger, Howard Kunreuther, James F. Lander, Eugene L. Lecomte, George Mader, Peter J. May, Michael J. McKee, Michael Michalek, William A. Mitchell, Elliott Mittler, Betty Hearn Morrow, John-Paul Mulilis, Mary Fran Myers, Eve Passerini, Dallas Peck, Ray Pena, Roy S. Popkin, E. L. Quarantelli, Richard J. Roth, Jr., Claire B. Rubin, James Russell, Stanley Schumm, Robert L. Schuster, Ellen Seidman, Frank H. Thomas, Kathleen Tierney, Susan K. Tubbesing, James L. Wescoat, Gilbert F. White, John D. Wiener, Ben Wisner, and Arthur Zeizel.

Appreciation is also extended to the people who reviewed and commented on the revised second review draft of the manuscript. These reviewers provided the sharp eyes needed to revise the manuscript into its final form. They brought a variety of perspectives to that review, including national policy, local government, natural science, social science, engineering, the private sector, and nongovernment organizations. Many thanks to the following second-draft reviewers:

Daniel Abrams, Michael Armstrong, Mary Carrido, Jack Cermak, Thomas Drabek, Nicholas Flores, William Hooke, Rocky Lopes, James Makris, Shirley Mattingly, Joanne M. Nigg, Ann Patton, James W. Russell, Randy Updike, and Gilbert F. White.

A very special "thank you" is due JoAnne DeRouen Darlington, who left Colorado State University to join the project staff at the University of Colorado at Boulder, to work as a postdoctoral fellow on this assessment. Her substantive contributions are greatly appreciated as well as her help in providing oversight and leadership for the graduate students who worked on the assessment.

Warm appreciation also goes to David Morton, who lent his patience and willingness to serve as a personal librarian to every contributor to the project.

Finally, literally thousands of pages of original draft text were edited as part of the process to produce this book. Jacquelyn L. Monday edited the manuscript pages in whatever form they arrived, without complaint. Jacki, you will always have my deepest appreciation for the excellent work that you provided as an editor and for your substantive contributions to the text as it emerged over several drafts. Mostly, thank you for taking this project on as your own and for the relentless commitment you brought to it.

Contents

DISASTERS by Design

Summary

A QUARTER-CENTURY AGO geographer Gilbert F. White and sociologist J. Eugene Haas published a pioneering report on the nation's ability to withstand and respond to natural disasters. At that time, research on disasters was dominated by physical scientists and engineers. As White and Haas pointed out in their *Assessment of Research on Natural Hazards*, little attempt had been made to tap the social sciences to better understand the economic, social, and political ramifications of extreme natural events.

White and Haas attempted to fill this void. But they also advanced the critical notion that rather than simply picking up the pieces after disasters, the nation could employ better planning, land-use controls, and other preventive and mitigation measures to reduce the toll in the first place. Today, at long last, public and private programs and policies have begun to adopt mitigation as the cornerstone of the nation's approach to addressing natural and technological hazards.

The 1975 report also had a profound impact by paving the way for an interdisciplinary approach to research and management, giving birth to a "hazards community"—people from many fields and agencies who address the myriad aspects of natural disasters. Hazards research now encompasses disciplines such as climatology, economics, engineering, geography, geology, law, meteorology, planning, seismology, and sociology. Professionals in those and other fields have continued to investigate how engineering projects, warnings, land use management, planning for response and recovery, insurance, and building codes can help individuals and groups adapt to natural hazards, as well as reduce the resulting deaths, injuries, costs, and social, environmental, and economic disruption. These dedicated people have greatly improved our understanding of the physical processes underlying natural hazards and the complexities of social decision making before, during, and after disasters. Yet troubling questions remain about why more progress has not been made in reducing dollar losses.

One central problem is that many of the accepted methods for coping with hazards have been based on the idea that people can use technology to control nature to make themselves safe. What's more, most strategies for managing hazards have followed a traditional planning model: study the problem, implement one solution, and move on to the next problem. This approach casts hazards as static and mitigation as an upward, positive, linear trend.

But events during the past quarter-century have shown that natural disasters and the technological hazards that may accompany them are not problems that can be solved in isolation. Rather, they are symptoms of broader and more basic problems. Losses from hazards—and the fact that the nation cannot seem to reduce them—result from shortsighted and narrow conceptions of the human relationship to the natural environment.

To redress those shortcomings, the nation must shift to a policy of "sustainable hazard mitigation." This concept links wise management of natural resources with local economic and social resiliency, viewing hazard mitigation as an integral part of a much larger context. Many aspects of this strategy were implicit in the recommendations formulated by White and Haas a quarter-century ago.

But to head off the continued rise in tolls from disasters, those principles must become more explicit.

This book reflects the efforts of over a hundred experts who have worked and debated since 1994 to take stock of Americans' relationship to hazards past, present, and—most importantly—future. Those contributions have been used to outline a comprehensive approach to enhancing society's ability to reduce the costs of disaster.

The Roots of the Problem

Many disaster losses—rather than stemming from unexpected events—are the predictable result of interactions among three major systems: the physical environment, which includes hazardous events; the social and demographic characteristics of the communities that experience them; and the buildings, roads, bridges, and other components of the constructed environment. Growing losses result partly from the fact that the nation's capital stock is expanding, but they also stem from the fact that all these systems—and their interactions—are becoming more complex with each passing year.

Three main influences are at work. First, the earth's physical systems are constantly changing—witness the current warming of the global climate. Scientists expect a warming climate to produce more dramatic meteorological events such as storms, floods, drought, and extreme temperatures. Second, recent and projected changes in the demographic composition and distribution of the U.S. population mean greater exposure to many hazards. The number of people residing in earthquake-prone regions and coastal counties subject to hurricanes, for example, is growing rapidly. Worsening inequality of wealth also makes many people more vulnerable to hazards and less able to recover from them. Third, the built environment—public utilities, transportation systems, communications, and homes and office buildings—is growing in density, making the potential losses from natural forces larger.

Settlement of hazardous areas has also destroyed local ecosystems that could have provided protection from natural perils. The draining of swamps in Florida and the bulldozing of steep hillsides for homes in California, for example, have disrupted natural runoff patterns and magnified flood hazards. And many mitigation efforts themselves degrade the environment and thus contribute to the next disaster. For example, levees built to provide flood protection can destroy riparian habitat and heighten downstream floods.

Another major problem has become clear over the past 20 years: some efforts to head off damages from natural hazards only postpone them. For example, communities below dams or behind levees may avoid losses from floods those structures were designed to prevent. But such communities often have more property to lose when those structures fail, because additional development occurred that counted on protection. Such a situation contributed to catastrophic damage from the 1993 floods in the Mississippi basin. And many of the nation's dams, bridges, and other structures are approaching the end of their designed life, revealing how little thought their backers and builders gave to events 50 years hence. Similarly, by providing advance warnings of severe storms, this country may well have encouraged more people to build in fragile coastal areas. Such development, in turn, makes the areas more vulnerable by destroying dunes and other protective natural features.

Fostering Local Sustainability

Sustainability means that a locality can tolerate—and overcome—damage, diminished productivity, and reduced quality of life from an extreme event without significant outside assistance. To achieve sustainability, communities must take responsibility for choosing where and how development proceeds. Toward that end, each locality evaluates its environmental resources and hazards, chooses future losses that it is will-

---

Disaster Losses Are Growing

From 1975 to 1994, natural hazards killed over 24,000 people and injured some 100,000 in the United States and its territories. About one-quarter of the deaths and half the injuries resulted from events that society would label as disasters. The rest resulted from less dramatic but more frequent events such as lightning strikes, car crashes owing to fog, and localized landslides.

The United States has succeeded in saving lives and reducing injuries from some natural hazards such as hurricanes over the last two decades. However, casualties from floods-the nation's most frequent and injurious natural hazard—have failed to decline substantially. And deaths from lightning and tornadoes have remained constant. Meanwhile injuries and deaths from dust storms, extreme cold, wildfire, and tropical storms have grown.

The dollar losses associated with most types of natural hazards are rising. A conservative estimate of total dollar losses during the past two decades is $500 billion (in 1994 dollars). More than 80 percent of these costs stemmed from climatological events, while around 10 percent resulted from earthquakes and volcanoes. Only 17

ing to bear, and ensures that development and other community actions and policies adhere to those goals.

Six objectives must simultaneously be reached to mitigate hazards in a sustainable way and stop the national trend toward increasing catastrophic losses from natural disasters.

- ***Maintain and enhance environmental quality.*** Human activities to mitigate hazards should not reduce the carrying capacity of the ecosystem, for doing so increases losses from hazards in the longer term.
- ***Maintain and enhance people's quality of life.*** A population's quality of life includes, among other factors, access to income, education, health care, housing, and employment, as well as protection from disaster. To become sustainable, local communities must consciously define the quality of life they want and select only those mitigation strategies that do not detract from any aspect of that vision.
- ***Foster local resiliency and responsibility.*** Resiliency to disasters means a locale can withstand an extreme natural event with a tolerable level of losses. It takes mitigation actions consistent with achieving that level of protection.

---

percent were insured. Determining losses with a higher degree of accuracy is impossible because the United States has not established a systematic reporting method or a single repository for the data. Further, these numbers do not include indirect costs such as downtime for businesses, lost employment, environmental damage, or emotional effects on victims. Most of these losses result from events too small to qualify for federal assistance, and most are not insured, so victims must bear the costs.

Seven of the ten most costly disasters—based on dollar losses—in U.S. history occurred between 1989 and 1994. In fact, since 1989 the nation has frequently entered periods in which losses from catastrophic natural disasters averaged about $1 billion per week. The dramatic increase in disaster losses is expected to continue.

Many of the harshest recent disasters could have been far worse: had Hurricane Andrew been slower and wetter or torn through downtown Miami, for example, it would have wreaked devastation even more profound than the damage it did inflict. And the most catastrophic likely events, including a great earthquake in the Los Angeles area, have not yet occurred. Such a disaster would cause up to 5,000 deaths, 15,000 serious injuries, and $250 billion in direct economic losses.

• ***Recognize that vibrant local economies are essential.*** Communities should take mitigation actions that foster a strong local economy rather than detract from one.

• ***Ensure inter- and intra-generational equity.*** A sustainable community selects mitigation activities that reduce hazards across all ethnic, racial, and income groups, and between genders equally, now and in the future. The costs of today's advances are not shifted onto later generations or less powerful groups.

• ***Adopt local consensus building.*** A sustainable community selects mitigation strategies that evolve from full participation among all public and private stakeholders. The participatory process itself may be as important as the outcome.

A long-term, comprehensive plan for averting disaster losses and encouraging sustainability offers a locality the opportunity to coordinate its goals and policies. A community can best forge such a plan by tapping businesses and residents as well as experts and government officials. And while actual planning and follow-through must occur at the local level, a great deal of impetus must come from above. Nothing short of strong leadership from state and federal governments will ensure that planning for sustainable hazard mitigation and development occurs.

Mitigation Tools

Over the past few decades an array of techniques and practices has evolved to reduce and cope with losses from hazards and disasters. These and other tools will be vital in pursuing sustainable hazard mitigation.

Who is at Risk

Research has shown that people are typically unaware of all the risks and choices they face. They plan only for the immediate future, overestimate their ability to cope when disaster strikes, and rely heavily on emergency relief.

Hazard researchers now also recognize that demographic differences play a large role in determining the risks people encounter, whether and how they prepare for disasters, and how they fare when disasters occur. For example, non-minorities and

Land Use

Wise land-use planning that limits expansion into sensitive areas is essential to sustainable hazard mitigation. Indeed, land-use planning, hazard mitigation, and sustainable communities are concepts with a shared vision in which people and property are kept out of the way of hazards, the mitigative qualities of the natural environment are maintained, and development is resilient in the face of natural forces.

Unfortunately, no overarching guidance informs development in hazard-prone areas. Instead, a patchwork of innumerable federal, state, and local regulations creates a confusing picture and often reduces short-term losses while allowing the potential for catastrophic losses to grow. This scattershot approach, as well as the federal and state trend to cut risk and assume liability, have undermined the responsibility of local governments for using land-use management techniques to reduce exposure to hazards.

Warnings

Since the first assessment was completed, significant improvements in short-term forecasts and warnings (hours to days ahead of a hazardous event) have dramatically reduced loss of life and injury in the United States. Yet many communities lag in their ability to provide citizens with effective warning messages. The nation needs to make local warning systems more uniform, develop a comprehensive model for how they work, and provide this information to local communities along with technical assistance. Better local management and decision making are now more critical than most future advances in technology.

---

households with higher socioeconomic status fare better, while low-income households are at greater risk mainly because they live in lower-quality housing, and because disasters exacerbate poverty.

The need for mitigation and response efforts that acknowledge the demographic differences among the nation's citizens will become even more critical as the U.S. population becomes more diverse. Research is also needed to shed further light on how mitigation programs ranging from public education to disaster relief can be rendered equitably.

It's also important to remember that short-term warning systems do not significantly limit damage to the built environment, nor do they mitigate economic disruption from disasters. Long-range forecasts that help define the risks to local communities years to decades ahead of potential hazards could assist local decisionmakers in designing their communities to endure them.

Engineering and building codes

The ability of the built environment to withstand the impacts of natural forces plays a direct role in determining the casualties and dollar costs of disasters. Disaster-resistant construction of buildings and infrastructure is therefore an essential component of local resiliency. Engineering codes, standards, and practices have been promulgated for natural hazards. Local governments have also traditionally enacted building codes. However, investigations after disasters have revealed shortcomings in construction techniques and code enforcement. Codes, standards, and practices for all hazards must be reevaluated in light of the goal of sustainable mitigation, and communities must improve adherence to them.

Insurance

The public increasingly looks to insurance to compensate for losses from many types of risk-taking behavior. However, most property owners do not buy coverage against special perils, notably earthquakes, hurricanes, and floods. For example, nationwide only about 20 percent of the homes exposed to floods are insured for them. Many people assume that federal disaster assistance will function as a kind of hazard insurance, but such aid is almost always limited. And even when larger amounts are available, they are usually offered in the form of loans, not outright grants.

Insurance does help minimize some disruption by ensuring that people with coverage receive compensation for their losses as they begin to recover. The insurance industry could facilitate mitigation by providing information and education, helping to create model codes, offering financial incentives that encourage mitigation, and limiting the availability of insurance in high-hazard areas.

The industry already has problems providing insurance in areas subject to catastrophic losses because many insurers do not have the resources to pay for a worst-case event. Furthermore, the current regulatory system makes it difficult to aggregate adequate capital to cover low-frequency but high-consequence events.

New technology

Computer-mediated communication systems, geographic information systems (GIS), remote sensing, electronic decision-support systems, and risk-analysis techniques have developed substantially during the last two decades and show great promise for supporting sustainable hazard mitigation. For example, GIS models enable managers to consolidate information from a range of disciplines, including the natural and social sciences and engineering, and to formulate plans accordingly.

Remote sensing can be used to make land-use maps and show changes over time, feed information to GIS models, and gather information in the wake of disasters. Finally, decision-support systems can fill a gap in hazards management by analyzing information from core databases, including data on building inventories, infrastructure, demographics, and risk. The systems can then be used to ask "what-if" questions about future losses to inform today's decision making. Such systems are now constrained by the lack of comprehensive local data, but they will become more important as the process of evaluating and managing risk grows in complexity.

Essential Steps

The shift to a sustainable approach to hazard mitigation will require extraordinary actions. Here are several essential steps; note that many initial efforts are already under way.

Build local networks, capability, and consensus

Today hazard specialists, emergency planners, resource managers, community planners, and other local stakeholders seek to solve problems on their own. An approach is needed to forge local consensus on disaster resiliency and nurture it through the complex challenges of planning and implementation.

One potential approach is a "sustainable hazard mitigation network" in each of the nation's communities that would engage in collaborative problem solving. Each network would produce an integrated, comprehensive plan linking land-use, environmental, social, and economic goals.

An effective plan would also identify hazards, estimate potential losses, and assess the region's environmental carrying capacity. The stakeholder network especially needs to determine the amount and kind of damage that those who experience disasters can bear. These plans would enable policymakers, businesses, and residents to understand the limitations of their region and work together to address them. Full consensus may never be reached, but the process is key because it can generate ideas and foster the sense of community required to mitigate hazards.

This kind of holistic approach will also situate mitigation in the context of other community goals that, historically, have worked against action to reduce hazards. Finally, the process will advance the idea that each locality controls the character of its disasters, forcing stakeholders to take responsibility for natural hazards and resources and realize that the decisions they make today will determine future losses.

Federal and state agencies could provide leadership in this process by sponsoring-through technical and financial support—a few prototype networks such as model communities or regional projects.

Establish a holistic government framework

To facilitate sustainable mitigation, all policies and programs related to hazards and sustainability should be integrated and consistent. One

Emergency Preparedness and Recovery

Even if encouraged by more holistic state and federal policies, sustainable hazard mitigation will never eliminate the need for plans to address the destruction and human suffering imposed by disasters. In fact, one way to progress toward sustainable hazard mitigation is by creating policies for disaster preparedness, response, and recovery that support that goal.

A great deal of research has focused on pre-disaster planning and response since the 1975 assessment. Studies have found that pre-disaster planning can save lives and injuries, limit property damage, and minimize disruptions, enabling communities to recover more quickly.

Recovery was once viewed as a linear phenomenon, with discrete stages and end products. Today it is seen as a process that entails decision making and interaction among all stakeholders—households, businesses, and the community at large. Repossible approach toward this goal is a conference or series of conferences that enable federal, state, county, and city officials to reexamine the statutory and regulatory foundations of hazard mitigation and preparedness, in light of the principles of sustainable mitigation. Potential changes include limiting the subsidization of risk, making better use of incentives, setting a federal policy for guiding land use, and fostering collaboration among agencies, nongovernmental organizations, and the private sector.

Other efforts to foster a comprehensive government framework could include a joint congressional committee hearing, a congressional report, a conference by the American Planning Association to review experiences in sample communities, and a joint meeting of federal, state, and professional research organizations.

Conduct a nationwide hazard and risk assessment

Not enough is known about the changes in or interactions among the physical, social, and constructed systems that are reshaping the nation's hazardous future. A national risk assessment should meld information from those three systems so hazards can be estimated interactively and comprehensively, to support local efforts on sustainable mitigation.

Local planning will require multi-hazard, community-scale risk as-

---

search has also shown that recovery is most effective when community-based organizations assume principal responsibility, supplemented by outside technical and financial assistance. An even further shift—away from an exclusive focus on restoring damaged structures toward effective decision making at all levels—may be needed. Outside technical assistance can help strengthen local organizational and decision-making capacity.

Local leaders too often fail to take advantage of the recovery period to reshape their devastated communities to withstand future events. Most local disaster plans need to be extended not only to explicitly address recovery and reconstruction but to identify opportunities for rebuilding in safer ways and in safer places.

Fortunately, revisions to disaster legislation in the last several years have allowed a greater percentage of federal relief monies to fund mitigation programs. Pre-disaster planning for post-disaster recovery is vital to communities' ability to become disaster resilient.

sessment maps that incorporate information ranging from global physical processes to local resources and buildings. This information is not now available, and will require federal investment in research on risk-analysis tools and dissemination to local governments.

Build national databases

The nation must collect, analyze, and store standardized data on losses from past and current disasters, thereby establishing a baseline for comparison with future losses. This database should include information on the types of losses, their locations, their specific causes, and the actual dollar amounts, taking into account problems of double-counting, comparisons with gross domestic product, and the distinction between regional and national impacts. A second database is needed to collate information on mitigation efforts—what they are, where they occur, and how much they cost—to provide a baseline for local cost-benefit analy-

A New Approach to Hazards

Researchers and practitioners in the hazards community need to shift their strategy to cope with the complex factors that contribute to disasters in today's—and especially tomorrow's—world. Here are the main guidelines for improving our ability to mitigate hazards.

• **Adopt a global systems perspective.** Rather than resulting from surprise environmental events, disasters arise from the interactions among the earth's physical systems, its human systems, and its built infrastructure. A broad view that encompasses all three of these dynamic systems and interactions among them can enable professionals to find better solutions.

• **Accept responsibility for hazards and disasters.** Human beings—not nature—are the cause of disaster losses, which stem from choices about where and how human development will proceed. Nor is there a final solution to natural hazards, since technology cannot make the world safe from ALL the forces of nature.

• **Anticipate ambiguity and change.** The view that hazards are relatively static has led to the false conclusion that any mitigation effort is desirable and will—in some vague way—reduce the grand total of future losses. In reality, change can occur quickly

sis. These archives are fundamental to informed decision making and should be accessible to the public.

A central repository for hazard-related social science data is also lacking. This third central archive would speed development of standards for collecting and analyzing information on the social aspects of hazards and disasters.

Provide comprehensive education and training

Today hazard managers are being called upon to tackle problems they have never before confronted, such as understanding complex physical and social systems, conducting sophisticated cost-benefit analyses, and offering long-term solutions. Education in hazard mitigation and preparedness should therefore expand to include interdisciplinary and holistic degree programs. Members of the higher education community will have to invent university-based programs that move away from

---

and nonlinearly. Human adaptation to hazards must become as dynamic as the problems presented by hazards themselves.

• **Reject short-term thinking.** Mitigation as frequently conceived is too shortsighted. In general, people have a cultural and economic predisposition to think primarily in the short term. Sustainable mitigation will require a longer-term view that takes into account the overall effect of mitigation efforts on this and future generations.

• **Account for social forces.** Societal factors, such as how people view both hazards and mitigation efforts or how the free market operates, play a critical role in determining which steps are actually taken, which are overlooked, and thus the extent of future disaster losses. Because such social forces are now known to be much more powerful than disaster specialists previously thought, growing understanding of physical systems and improved technology cannot suffice. To effectively address natural hazards, mitigation must become a basic social value.

• **Embrace sustainable development principles.** Disasters are more likely where unsustainable development occurs, and the converse is also true: disasters hinder movement toward sustainability because, for example, they degrade the environment and undercut the quality of life. Sustainable mitigation activities should strengthen a community's social, economic, and environmental resiliency, and vice versa.

traditional disciplines toward interdisciplinary education that solves the real-world problems entailed in linking hazards and sustainability. This will require not only new degree programs but also changes in the way institutions of higher education reward faculty, who now are encouraged to do theoretical work.

Measure progress

Baselines for measuring sustainability should be established now so the nation can gauge future progress. Interim goals for mitigation and other aspects of managing hazards should be set, and progress in reaching those goals regularly evaluated. This effort will require determining how to apply criteria such as disaster resiliency, environmental quality, intra- and inter-generational equity, quality of life, and economic vitality to the plans and programs of local communities.

Also important is evaluating hazard-mitigation efforts already in place before taking further steps in the same direction. For example, the National Flood Insurance Program, which combines insurance, incentives, and land-use and building standards, has existed for 30 years, yet its effectiveness has never been thoroughly appraised.

Each disaster yields new knowledge relevant to hazard mitigation and disaster response and recovery, yet no entity collects this information systematically, synthesizes it into a coherent body of knowledge, and evaluates the nation's progress in putting knowledge into practice. Systematic post-disaster audits, called for in the 1975 assessment by White and Haas, are still needed.

Share knowledge internationally

The United States must share knowledge and technology related to sustainable hazard mitigation with other nations, and be willing to learn from those nations as well. Both here and abroad, disaster experts also need to collaborate with development experts to address the root causes of vulnerability to hazards, including overgrazing, deforestation, poverty, and unplanned development. Disaster reduction should be an inherent part of everyday development processes, and international development projects must consider vulnerability to disaster.

The Key Role of the Hazards Community

To support sustainable hazard mitigation, researchers and practitioners need to ask new questions as well as continue to investigate traditional topics. Important efforts will include interdisciplinary research and education, and the development of local hazard assessments, computer-generated decision-making aids, and holistic government policies.

Future work must also focus on techniques for enlisting public and governmental support for making sustainable hazard mitigation a fundamental social value. Members of the hazards community will play a critical role in initiating the urgently needed nationwide conversation on attaining that goal.

CHAPTER ONE

A Sustainability Framework for Natural and Technological Hazards

THIS BOOK IS THE RESULT of the nation's second assessment of natural hazards, a multidisciplinary effort to assess, evaluate, and summarize knowledge about natural and related technological hazards and disasters from the perspectives of the physical, natural, social, and behavioral sciences and engineering. Technological hazards not related to natural events and disastrous events that result from violent human acts (e.g., terrorism and war) were excluded from the assessment. This volume summarizes the thinking of hundreds of nationally and internationally recognized experts who have worked and debated together since 1994 to take stock of this country's relationship to hazards in the past, the present, and, most importantly, the future.

It is clear from those deliberations and from the research carried out in past decades that natural and related technological hazards and disasters are not problems that can be solved in isolation. Rather, disasters and losses from hazards are symptoms of broader and more basic problems. A major thesis of this book is that losses from natural and related technological hazards—and the fact that the United States cannot seem to reduce them—are the consequence of narrow and short-sighted development patterns, cultural premises,[1] and attitudes toward both the natural environment and even science and technology. With every passing year the enviable physical, economic, and social circumstances of the United States are more vulnerable to natural and technological hazards. Through the premises, patterns, and processes that have been used to achieve this level of development, the United States has been—and still is—creating for itself increasingly catastrophic future disasters.

It is time for an evolutionary nationwide shift in the approach now being used for coping with natural and technological hazards by universally adopting goals that are broader than local loss reduction; by using a revised framework that links natural hazards to their global context, to environmental sustainability, and to social resiliency; and by modifying hazard mitigation efforts so that they are compatible with that new vision. This chapter explores briefly some of the historical roots of hazards management in the United States, gives an overview of the current hazards paradigm with its emphasis on reducing losses, and proposes a broader framework for future hazards research and management.

ORIGINS AND DEVELOPMENT OF THE CURRENT APPROACH

The Human Ecology School

Many of the disciplines that address the environment, hazards, and social interactions today had their origins in the school of thought known as human ecology, which was a developing subdiscipline of both sociology and geography at the University of Chicago at the turn of the last century. The human ecology perspective was philosophically explored by John Dewey, who wrote that the fact that humanity exists in a natural world that is innately hazardous results in human insecurity. Individuals and societies are thus compelled to seek security through the comfort of perceived absolute truths, such as religion, science, and philosophy (Dewey, 1929). More importantly, environmental perils such as floods and earthquakes do not exist independently of society because these perils are defined, reshaped, and redirected by human actions. Dewey's per-

[1] As used in this book, the term "culture" encompasses a wide range of characteristics of a society, including its economy, politics, technology, religion, and ideology.

spective was that "environmental problems stimulate inquiry and action, which transform the environment, engendering further problems, inquiries, actions, and consequences in a potentially endless chain" (Dewey, 1938, p. 28).

Dewey's ideas have been attributed with the profound distinction of having shaped a generation of social scientists who in turn shaped the young mind of geographer Gilbert F. White, internationally renowned today as the father of natural hazards research and management, while White was a student over 50 years ago at the University of Chicago (see Wescoat, 1992). White himself (1973) traces the origins of his ideas along a different path, but, like Dewey, he has consistently maintained that natural hazards are the result of interacting natural and social forces and that hazards and their impacts can be reduced through individual and social adjustment. In fact, it was White's pioneering dissertation that first asked the questions that still direct hazards inquiry today: Why are certain adjustments to hazards preferred over others? Why, despite investments in those adjustments, are social losses from hazards increasing? (White, 1945; White et al., 1958, 1997).

Subsequent investigations by the geographers who followed White have carried his perspective up to the present. The traditional geographer's approach has been to investigate how individuals and human collectives adapt and adjust to natural hazards by using "adjustments" such as land-use controls, protection works, and building codes and design standards.[2] The ultimate test of this research continues to be whether it succeeds in reducing such disaster losses as lives, injuries, and social and economic disruption (see Kates, 1962; Whyte, 1986; Burton et al., 1993; Mitchell, 1990; Palm, 1990).

The human ecology ideas that emerged at the University of Chicago also took hold in the young discipline of sociology. For example, MacIver (1931) explored the relationships between society and the natural environment, stating that "the revelation of the manner in which the environment molds and itself is modified by the life of the group is one of the chief achievements of sociology" (p. 308). Not unlike Dewey and White, MacIver saw adaptation as the physical, biological, and social adjust-

[2]Use of the terms "adjustment" and "adaptation" in this book generally follows the distinction made by Burton et al. (1978). That is, adjustments are purposeful or incidental and tend to be made over a short time frame. Adaptations are long-term cultural or biological coping mechanisms, often not the product of deliberate decisions on the part of the people who participate in them.

ment of society to the natural environment. He recognized that economic, political, and cultural factors combine with geographic ones in determining social modifications to the natural environment in ways that allow for the greatest fulfillment of human wants. Furthermore, he noted, the choice of "adjustment . . . implies valuation" (p. 309), so that some groups will inevitably end up with lower levels of satisfaction no matter what strategy is adopted.

MacIver's ideas were expanded by other sociologists who, like their counterparts in geography, sought to elaborate on society's adaptations to the natural environment. Unlike its counterpart in geography, the human ecology school in sociology had become nearly dormant by the 1960s. The natural hazards topic had been subsumed by sociology's more popular specialties—population studies and environmental sociology. A notable exception was Michael Micklin (1973), who offered a synthesizing view of societal adjustments by identifying adjustment mechanisms.

The Disaster Research School

An alternative current developed in sociology independently of the human ecological heritage. "Disaster research" began with Prince's (1920) dissertation on a technological disaster and was followed by investigations of natural disasters and inquiry into the nature and conditions of panic. Disaster research was fertilized during the 1950s by national anxiety over the Cold War. With federal funding, the National Research Council embarked on a series of investigations of disasters to learn lessons transferable to civil defense in the event of nuclear exchange with the Soviet Union. The specialty area, originally labeled "social disorganization," was based on expectations about what the research would discover. The findings from that research program have been synthesized in the social psychology of collective behavior and theories of social organization. (For a summary of the history of the field, see Quarantelli, 1995; for an assessment of the impacts of disaster research on social policy, see Dynes and Drabek, 1994; for a discussion of methods, see Stallings, 1997.)

Both theories are extraordinarily different from those of the human ecologists. The collective behaviorists offered explanations for human adjustment and behavior in the minutes, hours, and days after a disaster's impact. Scholars of social organization offered similar conclusions about the behavior of organizations. The "social disorganization" label for the research area was dropped as disasters were observed to strengthen rather than paralyze affected communities.

Specialized research topics beyond disaster impact studies have been pursued, such as warnings, short-term recovery, long-term community reconstruction, and social response to global climate change. Sociological research has traditionally emphasized reducing disaster losses by upgrading emergency preparedness for disaster response. Studies by the National Opinion Research Center established the importance of sending research teams to a stricken area immediately after a disaster occurred and also of making comparisons from one disaster to another. Subsequent investigations continued, and several attempts to summarize knowledge have been advanced (e.g., Drabek, 1986).

Blending Approaches and the Nation's First Assessment

By the 1970s, natural hazards research in geography (with its human ecological heritage and emphasis on loss reduction) and disaster research in sociology (with its collective behavior perspective and emphasis on disaster response and emergency preparedness) were both entrenched in their respective disciplines. Beginning in 1972, these approaches were mixed with the perspectives of climatology, economics, engineering, geology, law, meteorology, planning, psychology, public policy, and seismology. Geographer Gilbert White and sociologist Eugene Haas (1975), assisted by graduate students, scholars, practitioners, and policymakers, began the nation's first assessment of research on natural hazards—an effort to take stock of the nation's knowledge regarding natural hazards and disasters with an emphasis on the social sciences to suggest directions for national policy and to inventory future research needs.

That project lowered the walls that had separated many of the disciplines involved with natural hazards and paved the way for the interdisciplinary approaches to hazards research and management that the nation enjoys today. A new hazards paradigm did not emerge, but a shared and more integrated approach did. Planners incorporated a sociological perspective on human adjustment in the geographical/human ecological tradition; geographers conducted research on social response to disasters; seismologists learned the social psychology of risk communication; social scientists conducted risk analyses; engineers worked to understand the social science dimensions of technology transfer; and public administrators and policymakers drew on the knowledge of all disciplines.

The Hazards Adjustment Paradigm

Theoretical, empirical, and policy work on hazards has for the past several decades—and especially since the 1975 assessment—been based on the notion that human beings, as individuals and in groups such as families, communities, and businesses, choose how they cope with or adjust to extremes (hazards) in their environment. It defines natural hazards as extreme, low-probability meteorological or geological phenomena that have the potential to cause disasters when they strike human collectives. The paradigm uses the so-called bounded rationality model of individual decisionmaking (now adapted to cover groups, organizations, and businesses as well), which says that individuals make decisions based on limited knowledge and within constraints set by their social system. In this way they obtain satisfactory (though not necessarily optimal) outcomes. This decisionmaking model, coupled with the adjustment concept, generated the following five-step strategy for coping with hazards: (1) assess hazard vulnerability; (2) examine possible adjustments; (3) determine the human perception and estimation of the hazard; (4) analyze the decisionmaking process; and (5) identify the best adjustments, given social constraints, and evaluate their effectiveness (e.g., cost-benefit analyses; see Burton et al., 1978).

The public and private policies that have been developed based on this paradigm and model have generated a management strategy whose goal is to reduce hazard-related losses, such as lives, injuries, dollars, and social and economic disruption. The strategy is organized conceptually around a four-stage cycle that is described below. Implementation today relies on "loss reduction" activities in all four stages, fostered at the societal level but carried out locally or individually.

In current thinking, human adjustments to disasters take place throughout a cyclical process that has four stages: preparedness, response, recovery, and mitigation. These stages are rough but useful categories that have helped organize the thinking, activities, research, and policy for hazards management.

Preparedness involves building an emergency response and management capability before a disaster occurs to facilitate an effective response when needed. This requires a vulnerability analysis to identify what hazards could occur in a particular place and a risk analysis to determine the likely problems that an extreme event could impose. Preparedness also includes hazard detection and warning systems, identification of evacuation routes and shelters, maintenance of emergency supplies and communications systems, procedures for notifying and mobilizing key personnel, preestablished mutual aid agreements with neighboring communities, and other items. Also critical to preparedness are training for response personnel, conducting exercises and drills of emergency plans, and informing citizens through education programs.

Response refers to the actions taken immediately before, during, and after a disaster occurs to save lives, minimize damage to property, and enhance the effectiveness of recovery. The activities typically carried out during a response effort are hazard detection and warning, evacuation of threatened populations, shelter of victims, emergency medical care, search and rescue operations, and security and protection for property. Other activities can occur as well, depending on the disaster type, including the construction of temporary levees, closure of roads or bridges, provision of emergency water or power supplies, and attending to secondary hazards such as fire or the release of hazardous materials. The quality and timeliness of disaster response are typically functions of the planning and training done during the predisaster preparedness phase.

Disaster recovery involves short-term activities to restore vital support systems and long-term activities to return life to normal. An initial step in recovery is a comprehensive damage assessment to help set priorities. Recovery encompasses repairing and reconstructing houses, commercial establishments, public buildings, lifelines, and infrastructure; organizing and dealing with volunteers and donated goods; delivering disaster relief; restoring and coordinating vital community services; expediting permitting procedures; and coordinating activities among governments. Recovery can take a few weeks or several years, depending on a disaster's magnitude, the resources available to solve the problems, and other factors.

Finally, mitigation refers to the policies and activities that will reduce an area's vulnerability to damage from future disasters. These measures generally are in place before a disaster occurs. There are hundreds of mitigation actions that can be taken to reduce future damage and losses. In general, mitigation activities can be characterized as structural, infrastructural, and nonstructural.

Structural mitigation measures try to keep hazards away from people and buildings, or to strengthen buildings, infrastructure such as electrical power and transportation systems, or sites that are exposed to hazards. Levees, dams, and channel diversions are all examples of structural flood mitigation measures aimed at controlling the rate and flow of floodwaters so as to prevent damage. Building codes and construction practices are another form of structural mitigation since they can, for example, make buildings better able to withstand ground shaking in earthquakes.

Nonstructural mitigation measures attempt to distribute the population and the constructed environment such that their exposure to disaster losses is limited. Zoning ordinances can reduce exposure by limiting development or the density of human occupancy in particularly hazardous areas, by creating or maintaining open space, and by limiting the placement of critical facilities (hospitals, power plants, schools). Preferential taxation can provide incentives to maintain low density in hazardous areas.

Although insurance is a nonstructural adjustment, it does not directly reduce vulnerability to disasters. It does, however, relieve the financial burden on disaster victims and thereby taxpayers who may otherwise have to provide relief. Insurance helps spread the costs associated with disasters among a broad group of policyholders and can induce mitigation measures if premiums and/or deductibles reward such actions. Similarly, disaster relief programs of governmental and nongovernmental organizations lessen the impact of disasters on victims.

MOVING BEYOND EXISTING APPROACHES

The Need for Improvement

There is no reason to abandon the approach that has been used to date, and certainly the knowledge that has been accumulated through that research design is valid. Yet there are some troubling questions about why more progress does not appear to have been made in reducing national losses from hazards. The use of traditional adjustments to natural and related technological hazards has been extensive (though certainly not universal) throughout the United States and across hazard types. Some successes have been achieved (and are detailed in later chapters), but the overall situation is that (1) the already-staggering monetary losses from disasters are still increasing; (2) there is reason to believe that in many instances mitigation activities are simply postponing losses that will be more catastrophic when they do occur; and (3) many efforts at disaster mitigation and many disasters result in short-term or cumulative environmental degradation and ecological imbalance, which, besides being detrimental to society, also contributes to the occurrence and severity of the next disaster.

Increased Losses

Although much more is said in Chapter 3 about disaster losses and how they are estimated, it should be noted here that even today's most liberal estimates are probably far too low to reflect the true cost of disasters. Accuracy aside, virtually all estimates clearly demonstrate that losses continue to escalate.

In 1970 total U.S. direct losses from natural disasters were estimated at $4.5 billion annually. Today, estimates range from $6 billion to $10 billion annually; some claim the figure will reach $17 billion by the year 2000 (all in 1970 dollars). Still others claim that by including crop damage from hail and the impacts of extreme heat and cold the annual losses today would be $20 billion. These estimates do not include indirect losses such as downtime for businesses, lost employment, environmental impacts, or emotional effects on victims. At least one broader estimate puts U.S. losses since 1989 at $52 billion annually.

Postponing Catastrophic Losses

Speculations abound today that many of the mitigation activities that have been undertaken—some for decades—are not preventing damage but merely postponing it. If the postponement amounts to many years, the accrued losses could be enormous. It is worth emphasizing that eventually nature will provide an event more extreme than that anticipated in plans and designs.

For example, communities located below dams or alongside levees construct themselves as if there is no risk of flooding, yet floods can and do occur that exceed the design capacity of control structures. When such things happen, the technological protection no longer offers security, but it has already made a larger population vulnerable to even more damage than was originally anticipated. In some cases, early evidence suggests that the mitigation measure is working, but when it is viewed over the longer term to include natural events that exceed the capacities of what was designed for it could well be judged a mistake. This may be the case with some engineered solutions to earthquakes, which may reduce losses in small- and moderate-sized quakes but work indirectly to increase losses in great earthquakes.

Environmental Degradation

As knowledge of complex ecosystems grows, many questions have been raised about the wisdom of permitting people to settle in certain areas (including hazard-prone ones), about what level of risk is acceptable for potentially environmentally harmful accidents such as toxic spills, and about what unintended (and even unimagined) impacts some mitigation activities may have.

For example, by alleviating (through hurricane warnings) some of the danger associated with living near the ocean, the United States may well have inadvertently increased the concentration and number of people and structures on the nation's coasts. This development demolishes dunes and other natural features, increases pollution, and in general exceeds the carrying capacity of the fragile coastal areas. Beach nourishment, a mitigation technique often touted as eminently sensible because it "restores" a natural protective feature, may be more disruptive of the littoral system and shoreline than is realized. The adverse environmental impacts of flood control dams are well known.

Proposed Shifts in Thinking

This tale of woe need not end here, however. There is no lack of ideas about what can be done to improve the research and implementation record and attempt to achieve on-the-ground results that are more impressive than those heretofore realized. The existing adjustment paradigm needs to evolve—indeed, it probably already has evolved—to begin to cope with the complex factors that contribute to natural disasters in today's world and especially tomorrow's. Researchers and practitioners of the hazards community need to make some adjustments of their own. Six shifts in thinking about natural hazards are proposed below.

1. *Adopt a global systems perspective.* First, the idea of surprise environmental extremes perhaps needs to be broadened or further illuminated, given the complexity of natural ecosystems and the human systems that interact with them. Systems theory, first proposed in the 1940s by the biologist Ludwig von Bertalanffy, may be a good place to start. Many different interactions are possible both within and between earth and social systems. These are the interactions that determine exposure to natural hazards. For example, anthropogenic practices affecting the earth system are changing the climate by polluting the air or destroying rain forests. Causal relationships within a given system include the effect of

increased volcanic activity on global climate, and the impact of intergovernmental relations on community disaster preparedness. The examples described in Chapter 4 hint at the range of complexities involved in interactions between earth and social systems, within and across the global-to-local levels of human aggregation.

2. *Accept responsibility for hazards and disasters.* Human beings, not nature, are the cause of disaster losses. The choices that are made about where and how human development will proceed actually determine the losses that will be suffered in future disasters. There is no final solution for adapting to natural hazards; technology cannot make the world safe from all the forces of nature.

3. *Anticipate ambiguity, constant change, and surprise.* The world is increasingly complex and interconnected. It is characterized by growing sets of interrelated problems, and change can occur quickly and in nonlinear ways. Most of these systems, unfortunately, have been designed for a predictable world in which future strategies and goals could be known and plugged into an existing organization. Attempts to manage the hazardous nature of the world have followed a traditional planning model: study the problem, develop alternatives, choose one, and move on to the next problem. This model views hazards as relatively static and sees mitigation as an upward, positive, and linear trend in the sense that any mitigation is desirable and will somehow decrease the grand total of losses in the future.

Hazards mitigation must become a process fed by the continuous acquisition of different mixes of new knowledge from different fields. Human adaptation to hazards must become just as dynamic as the ever-changing problems presented by hazards themselves.

4. *Reject short-term thinking.* Although it is by far the broadest and most potentially effective of all the strategies available under the current paradigm, mitigation as it is now frequently conceived of is too shortsighted. Certainly some efforts have been made to establish "permanent" mitigation—riverine open space and preserved wetlands come to mind. But today as the end of the design life of many of the nation's dams approaches and the carrying capacity of the bridges that connect barriers to the mainland is exceeded by a burgeoning evacuating population, it is becoming apparent how little thought was actually given to what lay 50 or more years beyond the initial decisions. In general, people have a cultural and economic predisposition to think primarily in the short term.

5. *Take a broader, more generous view of social forces and their role in hazards and disasters.* The adjustment paradigm and accompanying model of human choice focus on human agency, sometimes to the exclusion of other factors. Although the "bounded rationality" model of human choice explicitly recognizes the existence of constraining social, political, and economic forces and cultural values, recognizing those boundaries apparently has not helped break through them to reduce losses. It is possible, in fact, that those forces are much more powerful than previously thought. That is the premise adopted by the structuralist and social vulnerability schools of thought, which analyze abstract structures of human society that, its proponents say, actually dictate human adjustment to hazards and in some cases even instigate and perpetuate a hazardous situation. Much of this research has focused on the hazard vulnerability of indigenous populations and their traditional mechanisms for adapting to the environment (see Baird et al., 1975; Hewitt, 1976; Burton et al., 1978; Wisner et al., 1976; Carr, 1982; Drabek, 1986; White and Haas, 1975; Watts, 1983; and Blaikie et al., 1994). A few investigations in this country have examined the impact on individual earthquake hazard mitigation of a combination of legal, institutional, and economic structures and historic changes in the political economy (see Palm, 1983; Marston, 1982).

In addition, the limited resources available at all levels of society interact with the low salience of hazards to severely restrict social adjustment. The decentralized American political system requires a long list of players for effective action. Motivating, coordinating, and maintaining adjustment activities among these actors are difficult. For example, restricting the way in which land is developed creates winners and losers, and in today's world those threatened with economic loss are quick to mount a legal challenge, perhaps based on flaws in a local government's regulatory authority, or on constitutional protections from public "taking" of private property.

Although the adjustment model has done much to guide investigations of human decisionmaking with regard to hazards and has led to some actions to influence it, it turns out that humans do not hold all of the cards after all. There simply are too many other factors of all kinds that contribute to losses from natural disaster. Some can perhaps be modified, but others are probably immutable, and it would be wise to begin to give more serious consideration to them.

6. *Embrace the principles of sustainable development.* The concept of "sustainability" was elevated to global importance in the late 1980s by the World Commission on Environment and Development. It was defined as "development that meets the needs of the present without compromising the ability of future generations to meet their own needs" (1987, p. 188). The commission stressed that sustainability includes inter- and intragenerational equity; that adequate standards of living for all people should be possible; and that economics, ecology, and social equity are inseparable.

A globally accepted, operationalized definition of sustainability has yet to be offered. At this point it is more a philosophical perspective than a scientific concept. Some of the questions that it asks are: What kind of lifestyle do people want and need? How should people live now so that future generations are not penalized? How many future generations should be taken into account? What is essential for an acceptable quality of life? At what point should growth be limited? What risks are people willing to take in their interactions with the environment? Under what circumstances should prevention be chosen versus permissible risk? What, precisely, should be sustained and for whom? Who, or what, will monitor and manage continued sustainability and in whose interest? How can varied sectors such as government and business work together to define common interests?

The answers to these questions will vary from place to place and over time. And answering them on a global scale would inevitably engender conflicts over such issues as the redistribution of wealth, the extent to which overdeveloped and underdeveloped nations will need to change, the reshaping of national values such as individualism and consumerism that are so basic to personal and national identities, the conscious implementation of collective values in realms where they were been previously considered absent (such as science and economics), and the selection of factors to include in cost-benefit analyses.

Even without a precise definition, working toward sustainable communities (and, eventually, regions, nations, and the world) can go hand in hand with reducing losses from disasters. Not only are disasters more likely to occur where unsustainable development has taken place, but also the occurrence of a disaster itself hinders movement toward sustainability because of its resulting environmental degradation, ecological imbalance, socioeconomic impacts, and lowered quality of life. Therefore, actions designed to help communities mitigate disasters in a sustainable fashion should strengthen overall sustainability and resiliency to other social, economic, and environmental problems and vice versa (see Pramanik, 1993). A set of principles for sustainable hazards mitigation is proposed below.

PRINCIPLES OF SUSTAINABLE HAZARDS MITIGATION

Continuing along the same research and management path of the past several decades will bring increased frustration for researchers, practitioners, and policymakers. A broader perspective is needed so that far more complexity in both natural and human systems can be taken into account. Any new approach must be compatible with those social, cultural, and economic forces that cannot readily be changed and must encourage modification of those that need alteration for the overall good. A principal goal must be to foster truly long-term mitigation and loss reduction and to avoid burdening future generations with unnecessary hazards.

Most of these ideas and the principles that follow are not new. Many, in fact, were implicit in the adjustment paradigm as originally formulated by White and others decades ago. Others have been explored in alternative streams of research both in the United States and elsewhere. But it is time now to consolidate these currents of thought, which have too often seemed unrelated, as well as reassess the goals and consider new ones; make more explicit the principles that underlie the framework, some of which have become more obvious or more urgent in recent years; adopt a broader perspective where needed; reiterate research and management approaches that have proven effective; and, above all, reaffirm commitment to cooperation toward improving the nation's ability to avoid, withstand, and minimize losses from natural hazards. This new framework embraces the idea of adjusting to the environment but, it is hoped, both broadens and strengthens the adjustment approach that the research and management communities have been following for the past few decades.

The goal of this new sustainable hazards mitigation is not just to reduce losses but also to build sustainable local communities, with an eye toward expanding that resiliency to nationwide and international spheres.[3] Under this framework, actions to reduce losses would be taken

[3]The words "local" and "community" are used loosely throughout this volume to mean the smallest unit (larger than a household) that would be appropriate for participatory decisionmaking about a hazard-related issue. In general, that would mean a town, small city, or county in the United States. As envisioned in the sustainable hazards mitigation

only when they are consistent with the other principles of sustainability (Geis and Kutzmark, 1995). Sustainable hazards mitigation has six essential components: environmental quality, quality of life, disaster resiliency, economic vitality, inter- and intragenerational equity, and a participatory process.

1. *Maintain and, if possible, enhance environmental quality.* The first guiding principle of sustainability is that human activities in a particular locale should not reduce the carrying capacity of the ecosystem for any of its inhabitants. Hazard mitigation activities should be linked to efforts to control and ultimately reverse environmental degradation by coupling hazard reduction efforts to natural resource management and environmental preservation. Practices that degrade natural environments must be replaced with ones that allow continuous renewal of natural systems into the indefinite future.

In the United States the settlement of hazardous areas has destroyed local ecosystems that could mitigate those hazards. Draining swamps in Florida or bulldozing steep hillsides in California for homesites are examples of human actions that expose more people to natural hazards while destroying natural systems that would have helped minimize flooding. Limiting human expansion into areas with environmentally sensitive land-use planning has much to contribute to sustainable hazards mitigation. Urban sprawl and the transportation systems that promote it (i.e., automobiles) have significant negative impacts on environmental quality while also exposing increasing numbers of people to hazards. Linking the preservation of environmental quality to disaster mitigation could mean developing transportation systems that are less subject to disruption by disaster and less destructive of natural systems.

2. *Maintain and, if possible, enhance people's quality of life.* A population's quality of life has many components: income, education, health care, housing, employment, legal rights and exposure to crime, pollution, disease, disaster, and other risks. It is difficult to say at which policy level (local, state, national, global) standards for quality of life

---

framework, communities would not necessarily correspond to political entities but may instead be based on hazards- and/or resource-related boundaries (e.g., a watershed or an earthquake zone). It is acknowledged that delineating such entities would be problematic.

Sheriff Ben Picou directed the rescue of 60 pigs from a barn roof at Kaskaskia Island, Illinois, during the Midwest floods of 1993. A barge took the pigs to dry land. Photograph by Jim Richardson/NGS Image Collection.

should be set. For purposes of sustainability the full range of stakeholders in local communities (e.g., government, business, and individuals) should begin to consciously define and plan for the quality of life they want and believe they can achieve for themselves and future generations.

In tackling this issue localities must consider the externalities that may be imposed on neighboring areas (e.g., building levees and thus deferring the hazard to downstream communities); the nation (allowing development in hazardous areas, with the nation's taxpayers picking up the bill after a disaster); or the global community (producing toxic waste that will be disposed of in developing countries). Coping with externalities may require regulation in the form of taxes or penalties. By the same token, although community responsibility and a certain degree of self-sufficiency (at least in the case of hazards and disasters) are to be commended, economic realities will require interrelationships between a locale and other places. These relationships should be encouraged, and incentives may be needed.

3. *Foster local resiliency to and responsibility for disasters.* Local resiliency with regard to disasters means that a locale is able to with-

stand an extreme natural event without suffering devastating losses, damage, diminished productivity, or quality of life and without a large amount of assistance from outside the community. The locality takes responsibility for recognizing its environmental resources and the environmental hazards to which it is prone and chooses a level of hazard that it thinks is appropriate to its circumstances.

The first step toward responsibility and resiliency is public awareness of local environmental problems, specific natural hazards, disasters, environmental sustainability, and how they all affect each person's safety and security. Hazards and disasters should be approached as integral parts of the much larger context of environmental issues. The measures used to achieve resiliency will vary greatly from place to place and will depend on the types of natural and related technological hazards that are present, the economic practices engaged in, and the social factors that influence the local population's vulnerability (e.g., gender, age, ethnicity, social class; see Burton et al., 1993; Blaikie et al., 1994).

Incorporating sustainable hazards mitigation criteria into all new development plans and projects would make mitigation an ongoing focus. The period of recovery after a disaster is a good time to begin building resiliency because disasters provide political opportunities that can be capitalized on by managers, planners, and citizen groups. Often this requires tapping local knowledge to produce mitigation strategies that match local conditions. The current emphasis of the federal government on mitigation and the provision of funds to pursue it can help communities build resiliency.

4. *Recognize that sustainable, vital local economies are essential*. A sustainable local economy is diversified and thus less easily disrupted by disasters. A sustainable economy does not simply shift its externalities onto another region or onto the oceans or atmosphere. Nor can a sustainable economy be predicated on unlimited population growth, high consumption, or dependence on nonrenewable resources. Regional, national, and international cooperation and mechanisms will be needed to ensure that costs are determined accurately and distributed fairly. Thus, there are immense political, cultural, and social barriers in a capitalistic system that must be confronted.

5. *Identify and ensure inter- and intragenerational equity*. Intergenerational equity means not precluding a future generation's opportunity for satisfying lives by exhausting resources in the present generation,

destroying necessary natural systems, or passing along unnecessary hazards. A sustainable economy would preserve resources and ecosystems so that the costs of today's lives are not shifted to subsequent generations. Because future generations are stakeholders in absentia, sustainable hazards mitigation would not shift costs and hazards forward in time without considering their implications and whether appropriate benefits would accompany them. Intergenerational equity is clouded by many scientific uncertainties. The current nuclear waste crisis in the United States is one example of how what seemed like a good idea to one generation's scientists and policymakers may saddle hundreds of future generations with a hazardous exposure. A further complication is that the future costs of the externalities of production cannot be reliably estimated and thus tend to be ignored when decisions are made.

Intragenerational equity means a fair distribution of society's resources and hazards across today's population—all regions, genders, ethnic groups, and cultures. In the context of hazard mitigation, intragenerational inequities often translate into unequal distribution of exposure to hazards; a simple example is that people with limited resources often occupy more hazardous structures because those places cost less.

6. *Adopt a consensus-building approach, starting at the local level.* Building consensus is a process of seeking wide participation among all of the people who have a stake in the outcome of the decision being pondered, identifying all possible concerns and issues, generating ideas for dealing with them, and reaching agreement about how they will be resolved and what steps will be taken. It is important to note that full consensus may never be reached and may not even be desirable. Furthermore, the final decision may well be to proceed along a previously established, traditional course for one or more of the issues (e.g., using regulations). What is important is that the participatory process be engaged in, for the information it generates and distributes, for the sense of community it can foster, for the ideas that grow out of it, and for the sense of ownership it creates.

At least three ingrained cultural obstacles to consensus building can be identified immediately. Capitalism (with its emphasis on competition), Western thought (with its emphasis on individualism), and democracy (with its majority-rules mindset that creates winners and losers) all combine to frustrate truly cooperative endeavors. Nevertheless, there is evidence that people in the United States are able to transcend some of these innate characteristics in favor of wider cooperation for the sake of certain issues (fortunately for the hazards community the environment is one such issue) and under certain circumstances. The bias in favor of the status quo is a real one, but the consensus-building process has shown that people can often move beyond it if they feel responsible and empowered.

CONCLUSION

Many people have worked for a long time to reduce vulnerability to losses from natural and related technological disasters in the United States; the present is filled with the fruits of their accomplishments: strong infrastructures, comparatively safe buildings, effective warning systems, and perhaps most important the use of interdisciplinary solutions to solve disaster-related problems. But the natural and related technological catastrophes of the next millennium are likely to be larger than any ever before experienced, simply because that is the future that past actions have created. Intensive development in hazardous areas has dramatically increased exposure. Too many of the accepted methods of coping with hazards have been shortsighted, postponing losses into the future rather than eliminating them. People have sought to control nature and to realize the fantasy of using technology to make themselves totally safe.

Concern about that future brings about the call herein for an evolutionary shift in the way the United States and its communities relate to the natural environment. Hazards and disasters are not problems that can be solved in isolation—rather, they are part and parcel of much broader circumstances and processes. The need to pursue an alternative path is not an idea unique to the authors of the second assessment. The seeds for a sustainable approach to hazards mitigation have been planted everywhere. James Lee Witt, director of the Federal Emergency Management Agency, has called for the promotion of actions that mitigate adverse effects before disasters occur. President Clinton has stated that the time has come to mount a nationwide effort to reduce the impact of disasters as well as their economic consequences. Researchers have called for a broader view of the disaster problem (see Kreps and Drabek, 1996) and even for a revolution in approach (see Quarantelli, 1998). Calls have been made for longer-term perspectives, too, even in the first assessment (White and Haas, 1975). The idea of sustainability is built into the mission statements of many federal agencies (see Sidebar 1.1). In the 1960s President Johnson endorsed actions to promote wise use of the nation's valley lands that are subject to flooding, opening the door to sustainable policy and theory regarding floods. House Document 465 (Task Force

SIDEBAR 1.1

Federal Agencies with Missions that Support Sustainability

Two major initiatives bring federal agencies together to address natural hazards and disasters. First, the Subcommittee on Natural Disaster Reduction, of the Committee on Environment and Natural Resources in the White House's Office of Science and Technology Policy, consists of representatives of 15 federal agencies with programs related to natural disaster reduction. The subcommittee coordinates federal efforts for the International Decade for Natural Disaster Reduction, and its main goal is to facilitate information exchange to help create a sustainable society that is resilient to natural disasters. Second, many agencies have hazards-related missions, and 27 different federal agencies have some role (as described in the Federal Response Plan) in responding to presidentially declared disasters.

Sustainability

A federal interagency vision for sustainable development that includes natural and technological hazards provides an opening for a national shift to sustainable hazards mitigation. Portions of the vision statements of a handful of agencies, below, show the extent to which a holistic perspective and the concept of sustainability have already been institutionalized.

Federal Emergency Management Agency. FEMA's vision is that of a national public that knows what to do before, during, and after a disaster to protect people, their families, homes, and businesses; structures located out of harm's way; government and private organizations with plans, resources, and training for disaster response; and community plans for recovery and reconstruction after a disaster. FEMA is committed to an integrated and sustainable approach to hazards.

National Oceanic and Atmospheric Administration. NOAA envisions a world in which decisions are coupled with a comprehensive understanding of the environment and environmental stewardship, assessment, and prediction help enhance economic prosperity, quality of life, and protection of lives and property.

U.S. Environmental Protection Agency. EPA has a vision of a world in which all people's actions value the environment and ensure sustainable environmental and economic goals and the natural balance of all living things.

U.S. Forest Service. The Forest Service's vision is caring for the land and serving people through a land ethic that promotes the sustainability of ecosystems by ensuring their health, diversity, and productivity.

National Aeronautics and Space Administration. NASA uses a holistic perspective for environmental and natural resources research and studies the earth as an integrated system and the effects of natural and human-induced changes on the global environment.

Hazards-Related Missions

The hazards-related missions of the 15 federal agencies of the Subcommittee on Natural Disaster Reduction are summarized below. They cover many of the principles of sustainability.

Centers for Disease Control and Prevention. The CDC responds to natural disasters, investigates human health effects and medical consequences, provides epidemiological and scientific support for disaster planning and response, and recommends ways to mitigate the health consequences of future disasters.

Department of Energy, Office of Environment, Safety, and Health. DOE researches and implements policies, standards, and practices to reduce the effects of hazards on buildings, hazardous materials facilities, and electrical transmission structures.

Department of Transportation. DOT is responsible for transportation safety improvements and continuity of transportation services in the public interest and for providing technical assistance and transportation funds to states and cities.

Federal Emergency Management Agency. FEMA provides leadership and support to reduce loss of life and property and to protect institutions from all types of hazards through a comprehensive, risk-based, all-hazards emergency management program of mitigation, preparedness, response, and recovery.

Federal Energy Regulatory Commission. FERC regulates all nonfederal public and private hydroelectric projects in the United States, including a dam safety program with engineering guidelines; flood methodology and criteria; seismic analyses and evaluations; and emergency plans for warnings, training, and public safety.

Federal Housing Administration. FHA administers programs to provide safe, affordable, and sanitary housing in a suitable living environment.

continued

on Federal Flood Control Policy, 1966) asserted that a basic national goal is to foster efficient use of the bottomlands for the common good and computing national costs and benefits for all possible alternative uses.

The chapters the follow lay the groundwork for approaching hazards research and management from the standpoint of sustainability. In the next chapter are three visions of ideal, sustainable cities. Next are summaries of the knowledge gained through research and experience in

SIDEBAR 1.1 *Continued*

National Institute of Standards and Technology. NIST Building and Fire Research Laboratory performs research and development to improve standards and practices for buildings and lifelines to reduce losses from earthquakes, extreme winds, and fire and to predict the behavior of fire and smoke and the performance of detection and suppression systems.

National Aeronautics and Space Administration. NASA obtains information from space about the earth to investigate processes of hazards; to provide space-based hazards mapping, risk assessment, and hazard monitoring; and to develop systems for information dissemination and hazards mitigation.

National Oceanic and Atmospheric Administration. NOAA, along with the National Weather Service, which it includes, describes and predicts changes in the earth's environment; promotes global stewardship of the oceans and atmosphere; and develops, maintains, and disseminates information on severe storms, flood warnings, weather forecasts, water resources forecasts, climate change predictions, and ocean and coastal analyses.

National Science Foundation. NSF sponsors and funds scientific and engineering research and education projects; supports cooperative multidisciplinary and interdisciplinary research in the United States and between the United States and other countries; and generates the knowledge necessary for better understanding, managing, and mitigating of natural disasters, including basic research on the physical processes that underlie hazards; impacts; prediction and warning; risk assessment and mitigation; disaster recovery and reconstruction; and social and behavioral responses.

the various aspects of hazards: losses, what influences people to mitigate hazards or ignore them, management and policy tools for sustainable hazards mitigation, disaster management, promising new ways of looking at hazards-related problems and coping with them, and specific steps and pieces of research that can and should be taken to further hazards mitigation efforts in this country.

Office of U.S. Foreign Disaster Assistance. OFDA coordinates the U.S. response to natural and technological disasters worldwide; provides assistance for emergency shelter, water, and sanitation; and promotes disaster prevention, mitigation, and preparedness.

U.S. Army Corps of Engineers. The Corps manages and executes engineering, construction, and real estate programs for federal agencies and foreign governments; supervises research and development; responds to natural emergencies; and provides information, technical services, and planning assistance on floods and floodplain issues.

U.S. Environmental Protection Agency. EPA improves and preserves the quality of the environment, both national and global; and protects human health and productivity of natural resources.

U.S. Forest Service. The Forest Service provides fire protection for life, property, and natural resources and technical assistance in training, prevention, and other areas.

U.S. Geological Survey. The USGS conducts research, transfers technology, and fosters policies and practices to reduce losses from earthquakes, volcanic eruptions, landslides, and hydrological hazards in the United States and abroad.

Other federal agencies are involved in some way with natural and related technological hazards, most with focused missions. For example, the Tennessee Valley Authority works to control flooding, replenish soil, promote improved agricultural practices, and produce electric power for the Tennessee River Valley. The National Institute of Mental Health, which is part of the National Institutes of Health, conducts and supports research on mental health and related services following disasters.

CHAPTER TWO

Scenarios of Sustainable Hazards Mitigation

THE 1975 NATIONAL ASSESSMENT of natural hazards (see White and Haas, 1975) presented various scenarios of hypothetical hurricane, flood, and earthquake disasters for the cities of Miami, Boulder, and San Francisco, respectively. The scenarios were written to show how large future U.S. disasters could be. They came uncannily close to foretelling the impacts and consequences of Hurricane Andrew in 1992, the catastrophic 1993 Midwest floods, and the 1989 Loma Prieta and 1994 Northridge earthquakes, which together have cost the nation some $100 billion.

Innovative mitigation activities have been developed and implemented since 1975 to lessen some communities' vulnerability to natural disaster. One consequence (see Chapter 3) has been that loss of life from U.S. disasters in the past two decades has gone down, although the average annual cost in dollars has continued to rise. In addition, the potential for catastrophic loss of life in extraordinarily large future events is now greater than ever.

Like the 1975 assessment, this chapter presents three scenarios. Unlike the 1975 scenarios, the ones here are not designed to illustrate how large future disasters could be. Instead, they were written to illustrate some of the previously ignored causes of disasters and to suggest how to tie together the related concepts of disaster resiliency, economic vitality, environmental quality, and quality of life. These scenarios are by no means comprehensive and focus briefly on only a few issues in each city: the natural environment and social capital in Miami, decisionmaking and education in Boulder, and energy and transportation in San Francisco. Of course, all of the perplexing problems involved in sustainable hazards mitigation have not been addressed in these brief scenarios, and more questions have probably been raised than answered. Some may say that the scenarios are fanciful and Pollyannaish, and to this the authors plead guilty. They are primarily fantasy, although the statistics and facts about the predisaster community are all true, and damage predictions are reasonable averages of those found in the hazards research literature. Despite the creative license taken with potential future social behavior, it is hoped that these scenarios prompt readers to think about and discuss the possibility of making sustainable hazards mitigation a reality.

MIAMI, FLORIDA

Perhaps because very few catastrophic hurricanes made landfall in the United States between 1965 and 1989, the focus of major national attention and programs was on other types of natural hazards, and hurricane loss potential was given relatively little attention. During this time, the eastern coast of the United States became densely developed and populated. South Florida grew from fewer than 3 million people in 1950 to over 13 million in 1991, and 80 percent of that growth was in coastal areas. After some especially damaging hurricanes in the early 1990s, some researchers noted that hurricane activity seemed to be returning to a cycle similar to that of the 1940s and 1950s.

After many years of disagreement and uncertainty, most climate scientists were in agreement that the earth's climate was changing. The highly respected Intergovernmental Panel on Climate Change came to an unambiguous conclusion in 1995. Global warming, it said, was occurring and continued emissions of greenhouse gases, through a direct chain of events, would cause a rise in sea level and increasingly malevolent weather extremes, such as flooding. Many policymakers ignored or contested the implications, but scientists were in agreement, and insurance and reinsurance companies studying the data became particularly worried. For example, the president of the Reinsurance Association of

America claimed that climate change could bankrupt the industry. In addition, throughout the years insurance agencies had been creatively "protecting" their assets from potential catastrophe, thus leaving south Florida in the dangerous position of possibly having no reliable property insurance in the aftermath of a large hurricane. Unsafe and unplanned development, climate change, and social institutions conspired to make Miami a hazardous place to live and work.

Hurricane Sirin

In the midst of national handwringing over the hurricane vulnerability of coastal communities, Hurricane Sirin made landfall. Official pre-Sirin estimates predicted that 41 to 99 hours would be needed to evacuate everyone in this part of Florida. However, 21 hours after the first evacuation notice, Hurricane Sirin hit Miami Beach as a category 4 storm. After destroying much of Miami, it moved across the Everglades and Big Cypress National Preserve, devastated much of the Fort Myers area on the western Florida coast, entered the Gulf of Mexico, flooded New Orleans after it was downgraded to a tropical storm, and caused rainstorms and flooding in Mississippi and Tennessee. Hurricane Sirin died out as a rainstorm in Kentucky. The joint efforts of the National Oceanic and Atmospheric Administration's National Weather Service, the National Hurricane Center in Miami, and Miami's vigilant local officials had provided one of the best hurricane forecast and warning systems in the world. But the warning system could not compensate for the consequences imposed by previous rapid environmental destruction and unsound and unplanned development. Hurricane Sirin surpassed all previous hurricanes in the numbers of people injured, killed, and left homeless and jobless and in environmental and property damage. Most critical facilities (water, phone, sewer, utilities, transportation, shelters, hospitals, fire and police stations) were either destroyed or damaged. Flooding and water seepage after roofs blew off did further damage to critical public and private property.

After Hurricane Andrew in 1992, Florida had led the nation in exploring the feasibility of working toward sustainability, including the worthwhile future of a state made up of disaster-resilient communities. By the time Hurricane Sirin struck, sustainability was not only on the tip of most officials' tongues, it was actually becoming a household word. Sirin created a window of opportunity for sustainable mitigation strategies during reconstruction. Because Florida already had many post-

Motorists battle the 145-mph winds of Hurricane Andrew, which struck south Florida on August 24, 1992. Photograph by C.J. Walker, copyright *Palm Beach Post*.

disaster recovery plans in place, and owing to the insight of a number of key figures (the governor, the state director of emergency services, the mayor of Miami, many other local officials, hazards professionals, the insurance industry, and a growing number of environmental justice and sustainable development experts), Miami and other affected areas reconstructed themselves as national models of sustainability and disaster resiliency. The region built on its strengths: a strategic gateway linking North America, South America, and the Caribbean; a multicultural, multilingual work force and community; warm weather; unique natural ecosystems; and a worldwide reputation as a tourist attraction.

Ecosystem Health

The main reason Hurricane Sirin resulted in the largest disaster in U.S. history was the way human settlements had developed on the south Florida coastline in the latter part of the twentieth century. Development density along the coastal ridge had reduced the ability of natural systems to provide protection from hurricanes. For example, area beaches, which had once served as natural protection from severe weather and storm surges (and provided habitat for endangered turtles), were of little use during Hurricane Sirin because houses, condominiums, and hotels with walls primarily of glass were located right on top of them—in fact, the beaches had been eroding because of excessive development before Hurricane Sirin struck.

In addition, natural stands of mangrove trees had been considered a nuisance and were removed with fervor. They once had stabilized the shorelines; filtered runoff; acted as a buffer against storm winds and tides; and provided a habitat for shrimp, fish, and crabs. Finally, since the early 1900s, a variety of local, state, and national laws and regulations had encouraged the draining, filling, channeling, damming, and general demise of the Florida Everglades. What was left of the Everglades—the nation's only tropical national park—supported a number of important ecosystems and purified south Florida's drinking water. Human encroachment into the Everglades had become an increasingly important problem, as flood protection and the area's freshwater supply depended on the Everglades' health.

Not only were the natural systems not available to help dampen the impacts of Hurricane Sirin, they were tragically affected by the storm. The native ecosystems of south Florida originally had evolved to be able to bounce back after hurricanes, which after all are a normal occurrence in the region. However, encroachment and urbanization had weakened and whittled them down to such a small size that the trauma of Hurricane Sirin further damaged many beyond repair.

After Sirin it became clear to many people that it was socially, fiscally, and ecologically irresponsible to rebuild without protecting valuable natural resources. This meant some difficult decisions. For example, after two years of contentious, even bitter, debate and civic soul searching (and some arm twisting), it was decided that Miami Beach should be re-created as a sustainable mixed-use area. Much of the land was bought for a fair price by the city, state, and federal governments and turned into a national seashore consisting largely of public recreation areas and a few public buildings: a hurricane museum and wind/sun shelters for beachcombers, swimmers, and picnickers.

When it became clear that no insurance company in the world would provide coverage to any building on the outer shore, many owners sold their property, and some individuals donated small parcels of land. Others took the chance and rebuilt small self-insured hotels, restaurants, and golf courses—but they built low and sturdy since they knew they alone were financially responsible for future damage. There were no residential buildings left on the island. No cars were allowed on much of the Miami peninsula, and an extensive transit system—a hurricane-resistant solar/electric train—became an efficient way to visit and tour the beach. Once again, the beach became a barrier that protected the greater Miami area from storms, and it also protected Biscayne Bay and Miami's freshwater supply. The coastline became a wildlife sanctuary and a world-class national ecological and recreation treasure that drew more tourists to the area than before Hurricane Sirin's strike. Because few hotels were left on the beach, new ones sprung up inland. The new hotels were built with hurricanes in mind, including tsunami lobbies; window shutters; and roofs, walls, and foundations that were tied together. Similar reconstruction decisions were made in other ecologically sensitive hurricane-damaged areas of south Florida.

Insurance agencies, along with the city of Miami, banks, financial institutions, and the construction industry, increased their advocacy of building codes and became wary of insuring highly hazardous coastal and floodprone properties. At the same time, the federal government instituted strict policies that made it much more difficult for insurers to renege on their responsibility to their customers.

In addition, insurers and governments recognized that investment in businesses that contributed to increased global warming was incompatible with their interests. Consequently, they began reducing their stock in oil and coal companies and reinvested in less carbon-intensive energy technologies. Although their "greenhouse politics" were criticized by some, the bottom line was that they saved themselves and their clients' money and contributed to a better quality of life for future communities. This move on the part of the city and insurance and reinsurance industries ultimately pushed fossil fuel industries to reinvest in alternative energy sources. In time, as other global industries and investors followed suit, carbon emissions decreased and weather extremes, including hurricanes, began to stabilize.

Social Capital

Another predisaster factor that contributed to the severity of Hurricane Sirin's impact was the socioeconomic profile of the local population. A wide income gap existed, with a large number of people unemployed, underemployed, homeless, or poor, and there had been growing poverty even before the hurricane, especially among the elderly. In Miami per-capita income and home ownership were both low before Hurricane

Sirin, and one in three people lived in poverty in 1990. In fact, 40 percent of all Floridians had low or very low incomes before Hurricane Sirin. Florida also had a large homeless population because of a lack of affordable housing, unemployment, poor wages, family disintegration, poverty, lack of education, alcoholism, drug abuse, mental illness, and migration and immigration. One-third of Florida's homeless people were families. Before Hurricane Sirin, the state could shelter only 17 percent of its homeless.

Hurricane Sirin destroyed 90 percent of the affected counties' mobile homes, a primary source of low-income housing. Most of the low-income housing was particularly hard hit because it was not built to current code and most was located in hazardous areas. More costly homes with multiple roof lines, large unprotected windows, and gabled roofs also were especially vulnerable. Many homes, in all price ranges, that would have sustained only minor water and roof damage had they been built to code instead suffered massive water and structural damage when their roofs blew off. In the aftermath of the storm, affordable housing became nonexistent, and what rental units there were rose dramatically in cost. The storm created a massive exodus from the area when people lost their homes and jobs. This created difficult-to-calculate service burdens for the places to which they moved. Additionally, the more affluent neighborhoods got more media attention after impact and received more immediate and abundant relief supplies. Domestic violence rose as the stress of recovery, homelessness, joblessness, and uncertainty began to sink in and tensions mounted.

Despite these problems, Hurricane Sirin also made it clear to most people that Miami had to change dramatically if it was to recover and prosper in the future. The largest disaster in U.S. history robbed Floridians of their false sense of safety and exposed an infrastructure, social structure, and way of life that made Floridians recognize their vulnerability. It was also the last straw for a nation that had been paying for ever more expensive and tragic natural disasters. A compassionate and farsighted Congress realized that Florida's losses were also national losses and that losses of this magnitude should be prevented in the future. Slowly, policymakers realized that solving the problems of disaster mitigation and recovery meant solving age-old social problems. This was done in a variety of innovative ways. City planners, for example, knew that land-use decisions strongly affect both the population exposed to natural disasters and residential patterns that impact the availability of affordable housing. Consequently, they decided to make drastic changes in rebuilding the city.

Affordable disaster-resistant housing became a priority. A variety of locations around the city were rezoned for affordable housing, sometimes at the expense of highly profitable development. Diverse housing types were encouraged on adjacent sites. Infill development in existing city centers that already had infrastructure and services was preferred over outward sprawl. Making communities more compact, with mixed-use development, transportation infrastructure, schools, homeless shelters, emergency shelters, senior centers, prisons, and landfills all located in the region, further reduced vulnerability to disasters. Although some neighborhoods slated for infill protested, they were assured that, contrary to popular assumptions, density did not have to mean the same thing as congestion. In fact, as the city focused on revitalizing urban areas, infilling actually improved most people's quality of life. Care was taken to reduce the impact of development on critical natural resources, to preserve freshwater supplies, and to reduce urban sprawl.

Low-income residents and landowners with few resources to retrofit to comply with new building codes were offered help from the professional engineering, science, and planning associations and from the city and state. An interdisciplinary group brought together land management experts, community leaders, engineers, geographers, social scientists, economics experts, and local advocacy groups to help manage reconstruction projects for the poor. Faced with bearing a higher proportion of the costs locally, stakeholders tended to cooperate in ways never seen before. Contractors, engineers, architects, and other professionals used profits generated by affluent segments of society to fund locally sponsored design projects for poorer neighborhoods. This approach, labeled Robin Hood mitigation, was facilitated by short-term state and federal tax incentives; it was called one of the greatest achievements in disaster mitigation history and won the President's Annual Award for Sustainability upon the recommendation of the director of the Federal Emergency Management Agency.

Miami also drew on lessons and experiences from around the world to design comprehensive and effective programs for providing youths and adults with education and employable skills, supporting working families with child care and recreational opportunities, lowering crime rates, weatherizing and hurricane retrofitting homes, and promoting community gardens and open space. The city began incentive programs to encourage employers to offer flexible work schedules, job sharing, child care, elder care, and telecommuting. Despite early complaints that this would cut into profits, businesses found that it actually increased them by enhancing worker productivity and consumer loyalty and by decreasing employee absenteeism and turnover. Soon no incentives were needed from the city. With less commuter traffic, energy and time were conserved and the environment benefited, as did people's quality of life. The extensive investment in the new mass transit solar/electric trains not only made city life more pleasant and using public transportation more efficient, the city also reduced its carbon dioxide and other greenhouse gas emissions by 50 percent.

Investments in social capital created a population able to participate in important ongoing policy decisions and in increasingly professional and technical job markets and better able to recover on its own after hurricane disasters. People felt that their children had brighter futures and that everyone had created a healthier and fairer society. Tourism based on the attractions of the improved environment increased, as people came to see and be a part of Florida's "sustainable city."

Miami's leaders also realized that, as much as it had improved its own infrastructure, the city did not exist in a bubble. In fact, international trade through the city was a significant component of the national and regional economies. The city also realized that whatever the short-term gains it was not in its best interest to encourage dependency and ecological plunder of less developed countries. It prohibited policies or economic measures by the city, region, or local industry that would cause ecological and/or economic devastation in developing countries. The city began a policy of encouraging stronger ties with Latin American, European, and Caribbean nations.

Broader Implications

In achieving their transformation to a model of sustainable development and disaster resiliency, the city of Miami and south Florida received a great deal of help from other communities and the federal government. Once the transformation was in place, south Florida became much less of a drain on national resources; it did not require as much disaster aid, welfare, or social services. South Florida also became a national and global ambassador for sustainability. Tourism, which depends on healthy beaches, coastal resources, parks, and preserves, increased. It also increased because the city became a more attractive place to visit because of public transportation and a lower crime rate. But it became a world tourist destination because of its commitment to sustainability. The citizens of the world came to south Florida to see the city of the future. By restructuring the community so that it was affordable and accessible, had high levels of economic health, environmental quality, and equity, and was much less vulnerable to disasters, south Florida also made itself into a magnet for global business.

BOULDER, COLORADO

The history of development in Boulder, Colorado, cannot be separated from the history of flooding on Boulder Creek. When the first miners and eventual residents of Boulder arrived at the foot of the Rocky Mountains in October 1858, they heard tales of catastrophic flooding from the Southern Arapaho tribe, which had frequented the region for 200 years. Floods were a common occurrence during the early years of the town's development, and the worst flood in Boulder's history happened in 1894. Above-average snowfall, rapid warming resulting in fast snowmelt, and significant rain can cause serious flash flooding. In fact, Boulder has been rated as the city in Colorado in which the most people are the most likely to experience a catastrophic flood. A flood in Boulder of 12,000 cubic feet per second (cfs) is often called the 100-year flood, leading to the mistaken belief by many Boulder residents that such a flood will occur only once in 100 years.

Since the great flood of 1894, Boulder has had a series of interesting developments in its struggle to come to terms with its flood hazard. While many reports and studies have been written and performed, only in the past 25 years have major changes in flood policy been made. In 1969 a 25-year flood occurred in Boulder that made some officials pay attention. More significantly, the 1976 Big Thompson flash flood, 40 miles north of Boulder, killed 139 people and was destructive enough to capture the attention of many Boulder policymakers.

These experiences, as well as the restrictions required by the National Flood Insurance Program and the regional flood control board (Urban Drainage and Flood Control District), contributed to improvements in flood mitigation. Other important developments in Boulder's flood hazard management policy included the adoption of a warning system, a high-risk property purchase plan, and a nationally recognized greenways project. While extensive in nature, the warning system has never been fully tested and leaves several segments of the population unwarned and exposed. For example, Boulder draws a significant tourist population nearly year-round, and its large university houses a diverse international student population in the floodplain. Boulder also hosts a large transient homeless population, many of whom regularly inhabit the floodplain. The warning system is likely to be less effective for these special populations than for most permanent Boulder residents.

The city's buyout policy has been successful in removing some residences from the floodplain. Although it has been an expensive and localized solution, it is a permanent one. Many at-risk buildings have been razed and the resulting land converted to playing fields for the local high school. The land was also designed to increase the flood conveyance of the property, thereby lowering the surrounding flood elevations. The greenways project is a multipurpose riparian park that contains a heavily used bike/walkway, parks throughout the city, a whitewater course, a restored stream habitat, fishing ponds, breakaway bridges, and, most importantly, a use of the floodplain that does not increase the exposure to property and life. On the other hand, the city of Boulder has taken significant chances in its own property management and owns many buildings in the floodplain.

The Great Flood

Summer had been slow in coming the year of the great Boulder Creek flood. Snowpack in the mountains was well above average and skiers were delighted, but rain persisted down on the front range and most folks were anxious for summer to begin. The three days before the flood were sunny and warm, and flooding was the last thing on most citizens' minds. City officials, though, had been nervous for several weeks because, as is usual for the late May/early June period, Barker Reservoir (18 miles above Boulder on Boulder Creek) was full, and water had been flowing over the dam spillway for weeks. Moisture levels were well above average owing to the rapidly melting snow, high temperatures, and rain, and there was no room for excess water. Under these conditions a stalled rain storm 10 miles upstream from sunny Boulder was all it took to produce a flash flood.

When the last great flood occurred in 1894, the hills in the canyon above town had been stripped of timber by miners and loggers, which contributed to the high runoff into the creek. Although trees had since grown back, increased housing, highways, and local roads offset some of the benefits of the reforestation. Such development caused water to run quickly off hillsides and roadways into the creek. Five inches of rain fell in less than four hours, with no sign of letting up. A creek that normally runs at 300 to 600 cfs, whose banks can handle 8,000 cfs during wet seasons, grew quickly to 12,500 cfs.

At first, officials worried about the appropriate time to sound the alarms. The time required to warn the public of a flash flood in Boulder is longer than the time it takes a flood to form and reach the city. Therefore, the decision to announce an imminent flood and sound a warning must be made before enough rain has fallen to be sure that a flood will occur. On the one hand, Boulder officials feared losing credibility by sounding an unnecessary alarm; on the other, they feared not sounding an alarm soon enough for people to evacuate.

They eventually did sound the alarm and 20 minutes later were sure they had made the right decision when reports from upstream indicated that a 12-foot wall of water had formed 9 miles up the canyon. Forty-five minutes later it arrived at the mouth of the canyon, inundating the city of Boulder. As the water careened down the canyon, many homes were ripped from their foundations and others were damaged beyond repair. The now-furious wall of water collected parts of trees and private-access bridges, debris from destroyed and damaged homes, yard equipment, cars, sheds, rocks, silt, and propane tanks full of gas.

Although many city bridges were designed to break away on one side in the face of water pressure from a flood, the bridge on Broadway Avenue was a historical landmark. It was a low bridge not designed to break away, and debris and water backed up behind it. Because of this, more damage occurred behind the bridge than had been predicted. Finally, the aging bridge gave way to the force and weight of the water and the debris behind it, sending a rush of concrete, debris, asphalt slabs, and more water than ever anticipated toward the high school and downtown neighborhoods. The water washed north and blanketed residential areas and parts of downtown. Hundreds of parked cars throughout town were picked up by the floodwaters, some to careen slowly and precariously down streets, others to be swept away by the stronger current and became half-ton torpedoes in the creek.

The flood resulted in 30 deaths and hundreds of injuries. Half the deaths occurred in cars when people tried to cross flooded intersections. None realized that 18 inches of moving water can lift a car and carry it away. Economic damage was assessed at over $300 million, but that did not take into account the losses that could not be quantified, such as aesthetics, the sense of safety that many residents had permanently lost, psychological well-being, lost wages, and environmental degradation.

Sustainable Decisionmaking Processes and Cooperation

There was a silver lining, though, to this disaster. The flood was an "extraordinary moment," a wake-up call during which the citizens of Boulder were suddenly provided with a collective memory of a flash flood. The flood sparked imaginations and concerns and created a population far more willing to make long-term decisions for redevelopment in the floodplain that would result in a more flood-resilient community, to improve the city's warning system, to provide full-blown public education about floods, and to create detailed preflood plans.

However, flood resiliency was just the tip of the iceberg; immediately after the flood, the city began an assertive plan to link its disaster mitigation goals with its goals of ecological, social, and economic sustainability. A number of well-established Boulder County environmental citizens' groups had been generating public dialogue on the issue of sustainability long before the flood occurred. After the flood, they were sufficiently organized, educated, and enthusiastic to become informal consultants to the city. When word got out that the city was serious about addressing hazards, equity, power, growth, energy use, and diversity simultaneously, a wide variety of other local individuals and organizations volunteered their time and expertise to the undertaking. The city also hired local and outside experts in risk assessment, environmental problem solving, consensus building, conflict resolution, and educational outreach. The city was aided in this endeavor by the state and federal governments' new focus on facilitating local community commitment and capacity for undertaking innovative, collaborative, self-initiated, self-directed projects to increase resiliency and sustainability.

The city of Boulder began by assessing its current condition in terms of hazards, quality of life, level of resource consumption, and carrying capacity. It also examined the potential alternatives to the status quo on a range of issues (beginning with flood and wildfire hazards, education, housing, transportation, growth, equity, and regional cooperation). Then there was an assessment of what was essential to the community, given current knowledge about its problems, risks, and alternatives. Each of these steps took time, not only because widespread public participation was important but also because it involved gathering, assimilating, and discussing a great deal of information. However, a new paradigm grew out of these initial assessments, as did strong, flexible local institutions with the ability to take action on issues affecting long-term sustainability.

Various levels of cooperation were built by this new approach to

Boulder's problems. Institutional ties and horizontal integration were developed among a variety of local organizations and between organizations and local experts (professors, researchers, consultants, managers), students, nonactive citizens, and businesses. Also, ties and trust were increased between local organizations and the state and federal agencies that helped the city implement its plans. Citizens' organizations became equal partners with governmental officials in the decisionmaking process.

The great flood and its resulting social disruption brought into stark relief the idea that social and environmental problems rarely respect political boundaries. Slowly, Boulder and the surrounding towns began working together on a variety of development, transportation, and educational projects. Instead of each town being responsible for small parts of a creek, parcel of land, highway, or social problem, regional groups began to pool resources and responsibility and began thinking in terms of watersheds, ecosystems, communities, and regional quality of life goals.

In time Boulder was able to make definitive plans about mitigation and sustainability goals, act on them, and continually monitor and evaluate the effectiveness of its actions. At first Boulder officials viewed the meetings at which their risks and alternatives were assessed as a way to accomplish a specific goal—for example, to find out what to do about the flood hazard. However, they soon discovered not only that the procedure had evolved into a beneficial social institution but that the continual process of publicly assessing alternatives, priorities, and risks should never end because the contexts, facts, capacities, and commitments change with time.

Boulder also found that examining alternative solutions to a wide variety of integrated community problems, with the participation of a wide variety of citizens, resulted in better, more appropriate plans, a higher degree of consensus, more public participation, enhanced communication, and a less polluted, less vulnerable, more efficiently run city. The city focused on long-term solutions to complex problems. It stopped trying to prioritize one risk and interest over another and instead focused on what it would take to create the kind of city and level of safety that residents desired in the long run.

It was only through the long process of citizen participation in discussions in which the assumptions, values, uncertainties, and tradeoffs were made explicit that the city was able to decide the extent to which the floodplain should be regulated. In most cases the city followed the precautionary principle, which advises that in the absence of absolute proof a strong hypothesis that catastrophic losses are possible should be enough to encourage conservative action. The floodplain was significantly widened, and the resulting watershed parks became catch basins for stormwater runoff—a means to mitigate flooding as well as air pollution, a wildlife corridor, a bird sanctuary, a place of serene beauty, hiking trails, and, in other cases, busy rollerblade parks, baseball diamonds, soccer fields, picnic areas, and community gardens. In addition, warning systems were improved through public education and by placing unobtrusive, highly reliable solar warning devices in neighborhoods, within earshot of most residents.

Education

The city began an aggressive educational program about floods for children in kindergarten through twelfth grade so that every local child would know about floodplain ecosystems, flood hazards, flood predictions, damage prevention, and warnings. Adults were educated by innovative, often interactive programs and information brochures sponsored by neighborhood organizations and city officials. A nationally hailed Interactive Urban Ecology Museum (with a focus on local social, economic, and natural systems, including a large interactive exhibit about floods) was built and staffed primarily by volunteers consisting of well-trained high school students, senior citizens, and scientists from the university and local branches of federal agencies. The museum attracted and educated tourists as well as locals and was a model for a movement to create similar local ecology museums in other cities.

In the process of citywide education about flooding, many people began to realize that, as long as humans did not make themselves vulnerable by developing the floodplain, floods had some very beneficial qualities! For example, flooding deposits rich silts that replenish topsoil in the floodplain (leading to bumper crops in the new community gardens of the floodplain). Large piles of silt deposited by the flood were redistributed and used for topsoil, construction, and landfills, and local fly fishers found that fish communities flourished as a result of the flooding.

The educational focus on community and water did not end with flooding and often spilled over into some interesting public policy. For example, in studying water, Boulderites also began to understand that they lived at the edge of the Colorado plains, an arid desert. Everyone was aware that in the summer there are regular water shortages in the city. Through the years, residents had also begun to hear more and more about water rights disputes, and the cost of bringing water to Boulder was increasing. In an effort to lower water use without lowering the quality of life, the city began by xeriscaping all of its own ornamental property. Gone were the out-of-control sprinklers watering the grassy medians (and the roads) at noon on a sunny day. Instead, beautiful rock gardens full of drought-resistant native plants, flowers, cacti, and herbs were designed. Seeing the possibilities and encouraged by the city, citizens soon got rid of their own water-intensive lawns in favor of elegant, functional, drought-resistant patios, decks, rock gardens, flower gardens, and other xeriscaped delights.

Another example of innovative education was the new disaster mitigation degree at the University of Colorado. This interdisciplinary degree brought together researchers, managers, and practitioners from engineering, physical and social sciences, and policy studies to teach broad, holistic, comprehensive solutions to natural disaster problems.

Fearful Afterthoughts

The great flood was, in fact, fairly moderate. It could have occurred in the middle of the night, when most people would not have heard the sirens or radio and television warnings. It was a 100-year flood, but it could have been a 500- or 1,000-year flood. The Big Thompson flash flood of 1976, 40 miles north of Boulder, occurred under similar conditions, raised the water to 31,200 cfs, and killed 139 people. Such a flood is also possible in Boulder. A large truck or other obstruction could have fallen into the creek, causing water to be diverted from the creek channel onto, for example, Canyon Boulevard, swallowing areas that were never considered to be in danger of flooding. Barker Dam could have broken. It also could have been much worse if Boulder had not been as serious about flood mitigation as it had been for the few decades before the flood. In addition, Boulder has mostly high-technology industry, a high tax base, and a population with a higher than average education level—all of which probably helped in both pre- and postdisaster responses. Finally, of course, it could have been better. Much of the human life, social anguish, and monetary losses could have been minimized if more attention had been paid before the flood to floodplain management and sustainable development.

SAN FRANCISCO, CALIFORNIA

San Francisco is one of the most publicized metropolitan earthquake risk zones in the world, largely because of the 1906 earthquake there—an 8.25-magnitude event on a segment of the San Andreas fault. After that quake, fires destroyed three-fifths of the city's housing and the entire business district, and thousands were left dead, injured, or homeless. While the downtown business district was well insured, some of the municipal bonds used to reconstruct public works were not paid off until the 1980s. A repeat of the 1906 event is estimated by seismologists to have a 2 percent chance of occurring between 1990 and 2020. An event of magnitude 7.0 on the San Andreas fault is estimated to have a 23 percent chance of occurring during the same period. But the combined probability of a major earthquake occurring before 2020 along one of the faults in the San Francisco Bay Area is estimated to be 67 percent.

National and Local Shifts in Policy

By the late 1990s, natural disasters were beginning to be redefined as community-based problems that needed community-based solutions rather than unfortunate technological accidents or acts of God and nature. The nation's attention turned to disaster-resilient communities and sustainability. Many communities found innovative ways to mitigate hazards by reevaluating growth, land management, risk equity, and financial decisions; some, such as San Francisco, did such things before disaster struck.

The federal government played a key role in leading the nation's communities in this direction. Skyrocketing disaster recovery costs from increasingly frequent catastrophic events, combined with innovative leadership, helped agency leaders and Congress realize the many ways that the national government had historically encouraged unsustainability. They became unwilling to continue to foster unsustainable development in the nation's hazardous areas. The federal government took the bold bipartisan step of ending all subsidies to unsustainable endeavors (i.e., nonrenewable energy, building in hazardous areas, etc.), and began developing incentives that would encourage sustainable choices. In addition, the President spoke bluntly, noting that the federal government was a resource for techniques, expertise, and financial support for redesigning the nation's infrastructure but was not a bank with an infinite credit line to underwrite unwise local decisions. The President promised federal assistance in achieving a national sustainability initiative, akin to the federal effort to "put a man on the moon."

Federal and state governments began building a shared base of knowledge about sustainability and the long-term consequences of development and hazard and supported community projects to empower locals to make the needed cultural shift and to develop consensus. They found ways to disseminate technical information effectively and made plans to increase local commitment to and capacity for moving toward sustainability. Policies related to all hazards were integrated with policies for economic, social, and environmental objectives. Local communities and locally based branches of state and federal agencies were encouraged to develop innovative integrated plans for public investment and local resiliency. Locales with well-known disasters in their past were quick to seek solutions because new federal policies limited federal disaster relief to humanitarian assistance if the stricken locale had no sustainable hazards mitigation plan.

The idea of integrated policies and sustainability (e.g., policy on housing density had to be linked to policies on urban transportation, social equity, environmental quality, and disaster safety) was not new to the city and county of San Francisco or to many of its citizens. Longstanding concern for the city's earthquake safety, the President's pronouncements, and leadership from the recently installed director for the sustainable mitigation directorate of the Federal Emergency Management Agency led San Francisco to shift from many small-scale, largely experimental sustainability projects to one integrated, politically powerful, regional investment in sustainability ideals and practices. San Francisco formed a partnership with other bay-area counties, private-sector stakeholders, technical and scientific experts, and federal and state agencies to plan and make decisions about the area's future. The aim was a community that was more economically vibrant, disaster resilient, equitable, environmentally sound, and livable.

The city's first concern was to increase resiliency and quality of life for all citizens. Second, short-term gains that ignored long-term impacts and costs would no longer be tolerated. All city and regional policies were reviewed and revised to ensure that they did not subsidize or encourage unsustainability. The new focus of development policy became the creation of healthy communities in the long term, rather than short-term economic growth for its own sake. The chief concerns became education; affordable housing; transportation; energy conservation; creating resilient social, economic, and energy systems; and environmental quality.

Efforts were focused on accounting for the environmental and social costs of growth that were once hidden or externalized. Future resiliency against quakes—as well as landslides, floods, and fires—was a fundamental consideration in all policy decisions.

In keeping with the requirements of the new federal Urban Environment Act, the city created a Commission on the Environment to advise city boards; provide a centralized point of contact for environmental issues; regulate and enforce environmental standards; serve as a clearinghouse on sustainability projects; promote scientific, technical, and commercial exchanges; and help create and review sustainability goals. Some of these goals were to have by the year 2025 a 35 percent reduction in energy use, 75 percent of all waste recycled, and a 30 percent reduction in direct and indirect earthquake losses as projected by the city's three computer-generated models.

No one pretended that the solutions were easy. In most cases the city tried not to enforce changes without the deep support of local constituencies. There were many conflicts. In fact, major local and regional issues went unresolved for decades. But local, state, and federal initiatives continued to invest in the process of fair and informed negotiation, without losing sight of the goal of a sustainable future.

Resilient Energy Systems

One of the first major projects that San Francisco took on was lessening its reliance on complex and highly centralized utility grids. It learned a lesson from the 1994 Northridge earthquake when 3.1 million customers lost electricity and close to 100,000 homes and businesses were without power for over 24 hours. San Francisco began supplementing or replacing utility-grid energy with localized solar, wind, biomass, and hydroelectric systems. Decentralized renewable energy systems are less vulnerable to large power outages and make a community more self-reliant after disasters. They also lessen the chances of loss of power because of blackouts, terrorism, malfunctions, or fuel embargoes.

At first, some people balked at the initial expense of investing in disaster-resistant energy technologies. However, when the long-term costs of traditional energy were taken into account (including the cost of business disruptions following disasters, transmission and distribution, and environmental damage) and when the often-hidden benefits of the renewable energy system became part of the calculation (prevention of losses, shorter recovery time after disasters, lower operation costs, cleaner air) the critics realized that the cost of renewable energy was competitive. Adding to the competitiveness were new federal, state, and local policies that put an end to incentives that had steered development, financing, design, and construction of commercial buildings away from energy conservation, quake resiliency, and contributions to quality of life. All new buildings were required to use some form of self-sufficient, renewable energy system, and owners of older buildings were encouraged to retrofit.

General energy efficiency was also encouraged. Energy efficient windows—which are less likely to break in earthquakes—were installed in many buildings. Some innovative experiments were tried, including a large office building downtown that was redesigned so that excess heat from computers, lights, and air conditioning was redirected to a rooftop greenhouse. Organic vegetables were tended by volunteers and sold year-round to office workers on Fridays and to local residents every Saturday. Other downtown businesses formed partnerships to exchange their byproducts locally so that they could profit from each other's waste. City decisionmakers also tapped the job creation potential of the technologies used to implement environmental improvements: efficient lighting fixtures, photovoltaic panels, solar cars, and fuel cells. With partners in Alameda County, they encouraged manufacturing and research firms to locate in Oakland and subsidized the sharing of old design prototypes with nonprofit organizations throughout the bay area.

Resilient Transportation Systems

Transportation was given close attention. It was agreed that housing, work, shopping, and recreational opportunities should be near each other and that inexpensive efficient transportation should link them. After studying the costs of various options, it was concluded that contrary to popular opinion the cost of transportation could be significantly reduced if communities were redesigned to give more choices to citizens—bicycle paths, public transportation, and solar/electric cars. For example, a study in the 1990s that compared household automobile costs between a pedestrian-friendly section of northeast San Francisco and a more car-dependent suburban area found an annual difference of $13,000 between the two.

City leaders concluded that rail systems, bikes, and smaller solar/electric cars were a more sustainable form of transportation in earthquake country, that they were safer than personal cars, and that they reduced air pollution. With the help of a progressive gasoline tax the city

The 1994 earthquake in Northridge, California ruptured gas mains, sparking numerous fires. Photograph by David Butow.

embarked on an ambitious transportation plan, including a fleet of solar/electric city buses, hundreds of miles of bike and pedestrian paths, and an earthquake-proof light rail system (the city's cable car system, of course, was left as is). Not only did the changes in transportation free people from sitting in traffic, reduce traffic-related accidents (to car occupants and pedestrians), improve air quality, and increase neighborhood identity, it also lowered the chances of business losses in the aftermath of a disaster and ensured that residents would have access to everyday necessities in the aftermath of a disaster without being endangered or inconvenienced.

Related to transportation, a unified effort of the bay area counties also helped end urban sprawl and protect, once and for all, the bay area's Greenbelt, a broad band of open lands—3.75 million public and private acres—that surrounds the cities and towns of the bay area. It is one of the largest and most productive systems of open space in any metropolitan area, providing recreation, plant and wildlife habitat, clean air and water, community identity, and agricultural output worth $1.2 billion annually. In the 1990s it was estimated that much of the land would be developed by the year 2020.

Urban sprawl historically has been driven by the perception that costs are lower in outlying areas and that new communities do not have the problems (high taxes, crime, overcrowding) associated with urban areas. However, the costs of sprawl itself have often been unrecognized, deferred, externalized, and hard to quantify. In San Francisco, sprawl was creating a burden to be borne by future generations, by the environment, and by those who remained in the urban core. The cities and counties of the bay area successfully lobbied for changes in federal and state tax and transportation policies that had provided incentives for sprawl. By revitalizing the urban core, communities met their needs by building on a fraction of the land. Consequently, much of the Greenbelt was still intact in 2020. The importance of "green space" was also incorporated into city planning. If developers, for example, were unable or unwilling to incorporate parks into their developments, they were taxed to ensure that parks could be built nearby. This ensured the existence of aesthetic natural areas for recreation and wildlife and would also serve as fire breaks after a future earthquake and provide places for neighbors to camp, close to their homes, during aftershocks.

Community Prosperity and Resiliency

Of course, the previously mentioned changes in energy systems and transportation were just two of the many changes that occurred in San Francisco. One of the most dramatic overarching changes was in economic priorities. While considering and debating long-term costs and benefits, people realized that many current indicators of community and economic well-being were inadequate and also inappropriate to sustainability. For example, the traditional measures of prosperity—economic growth and employment levels—do not reflect people's lives and aspirations and, in fact, increases in such indicators often reflect human tragedy. The economy and employment levels often grow in the short term as a result of devastating disasters, the destruction of old-growth forests or aquatic ecosystems, increases in pollution, layoffs of professional workers, shoddily made appliances that must be replaced frequently, and the replacement of low-income housing with urban penthouses.

San Francisco wanted a broader definition of prosperity that included the value of natural resources, environmental protection, scenic beauty, education, literacy, safety, access to public lands, satisfaction with employment, health, earthquake resiliency, and the economic and social well-being of diverse future generations.

One result of this new way of thinking was new federal policies, which began transferring costs traditionally covered by federal tax dollars and private citizens to businesses. These long-recognized indirect costs included ecological degradation, potential costs from hazards, and large-scale unemployment. The incentives for self-insurance, private insurance, and managing a comprehensive "risk portfolio" were funded through federal tax incentives and regional bond issues. The transition was deliberately slow and gradual to avoid economic turmoil. An unprecedented long-range plan shifted actual risks and associated costs to those producing the risks and costs.

While San Francisco and the bay area were forging their long-term sustainability, other West Coast cities continued to decay from within; let unplanned and unsustainable growth continue; let the few take or spoil the social, ecological, and economic resources that belonged to the community; and ignored the consequences of poverty, vulnerability, and ecological ruin. San Francisco had become a strong, disaster-resilient community, capable not only of providing most of its citizens with a safe, high-quality way of life with little ecological impact but also of exporting sustainable values, techniques, know-how, and technology to other interested cities around the world. And then the ground under San Francisco began to shake violently. . . .

SUSTAINABILITY AND HAZARDS MITIGATION

The three scenarios in this chapter are a request to stop viewing natural disasters as isolated problems. The Miami scenario focuses on two root causes of disaster: historical ecological destruction and lack of investment in social capital. The Boulder scenario focuses on two ways of addressing the process of change: innovative decisionmaking and education. The San Francisco story focuses on creating more resilient lifelines by looking at two systems: energy and transportation.

No one would argue that a transition toward sustainable hazards mitigation is easy or that any one person has all of the answers. True sustainability is both a process and a goal, and many of the most important changes that need to be made are not immediately fathomable. The process of transforming the future requires open-minded debate; full public participation; a willingness to experiment, learn, fine-tune, and alter; and a consensus among stakeholders to stand behind their shared commitment to the goal.

Many disaster vulnerabilities can be addressed with existing tools and information, given a commitment to solving the problem. Disaster-resilient communities are built with the same building blocks that create resiliency to other social and environmental problems. Until people are ready to address the interdependent root causes of disasters and to do the difficult work of coming to negotiated consensus about which losses are acceptable, which are unacceptable, and what type of action to take, our nation's communities will continue on a path toward ever-larger natural catastrophes.

CHAPTER THREE

Losses, Costs, and Impacts

SUSTAINABLE HAZARDS MITIGATION requires local acceptance that the losses, costs, and impacts of future natural disasters will be the consequence of today's mitigation decisions (and nondecisions). This is a shift from current thinking and will require clear knowledge about likely local hazards and their losses, costs, and impacts. The same is true for the nation as a whole. Future loss estimation is needed, but the techniques and models in hand today are insufficient to make the difference in the shift to sustainable hazards mitigation. Furthermore, estimates of the losses, costs, and impacts of disasters that have already occurred are poor.

In an attempt to determine what these events have cost the nation, this chapter summarizes data on the losses, costs, and impacts of natural and related technological hazards and disasters in the two decades since the first assessment (White and Haas, 1975). The first part of the chapter catalogs deaths, injuries, and monetary losses from natural and technological hazards in the United States since 1975. Some of the more indirect and unquantifiable impacts of disasters are described and explained. Problems with the methods currently in use for measuring damage and other losses are presented. The chapter concludes with lessons that recent disaster losses provide about sustainability and with a call for a comprehensive, accurate, national database of disaster costs and mitigation measures.

DEATHS, INJURIES, AND DOLLAR LOSSES

It is estimated that natural hazards killed over 24,000 people (about 24 a week) and injured at least four times that many in the United States and its territories between January 1, 1975, and December 31, 1994. Of these, about one-quarter of the deaths and about one-half of the injuries occurred in natural disasters. The rest were the result of less dramatic but more frequent events, such as lightning strikes, car crashes owing to fog, and localized landslides.

During the same two decades, dollar losses to property and crops from natural hazards and disasters were between \$230 billion and \$1 trillion (in 1994 dollars standardized on the basis of the Consumer Price Index). A conservative estimate of the actual average dollar losses from natural hazards and disasters in the nation from 1975 to 1994 is \$500 billion, or about \$0.5 billion each week. These standardized dollar costs for the years 1975 to 1994 are shown in Figure 3.1.

Of these costs, more than 80 percent were imposed by climatological disasters, and about 10 percent were the result of earthquakes and volcanoes. Some 17 percent (or \$85 billion) were insured (Property Claim Services, 1996). From 1989 through 1994, about \$20 billion in losses were covered by presidentially declared disasters (Federal Emergency Management Agency, 1995). Nationwide distributions of presidential declarations by county from 1975 to 1994 are presented in Figure 3.2.

Since 1989 the nation has frequently entered periods in which losses from catastrophic natural disasters increased exponentially, averaging \$1 billion per week. However, most of the financial losses experienced in the period 1975 to 1994 occurred in events too small to qualify for federal assistance; most of the financial costs were not insured and thus losses were borne by the victims. The exponential increase in disaster losses is expected to continue (see Chapter 4). Munich Reinsurance (1995), the world's biggest insurance company, has warned that natural disasters are becoming costlier and apparently more frequent; it reports a record 600 disasters worldwide in 1996, up from 577 in 1995, with storms and floods accounting for over half.

To try to quantify the losses and costs of hazards and disasters in the

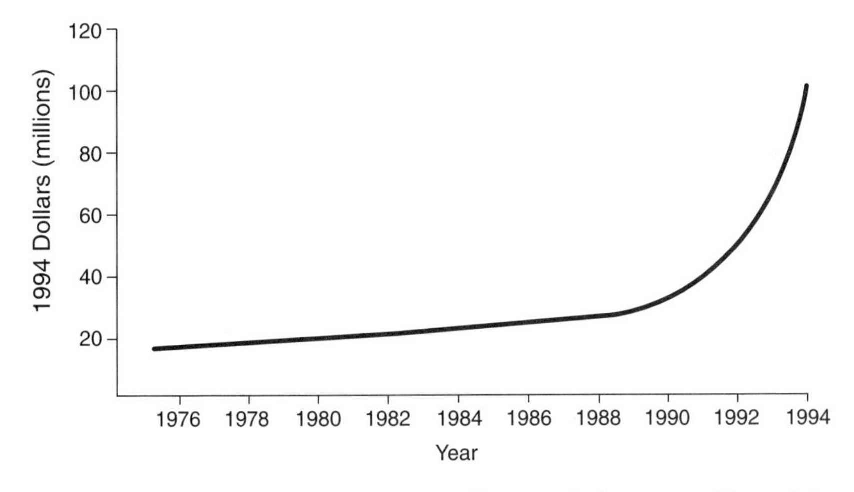

FIGURE 3.1 Average annual losses per 1 million people from natural hazards in the United States, 1975-1994 (in 1994 dollars).

United States, data were assembled on the money lost, deaths, and injuries from natural disasters during the 20-year period from 1975 to 1994. Loss data were collected on droughts, dust storms, earthquakes, extreme cold, floods, fog, heat, hurricanes, landslides, lightning, hail storms, ice/sleet, snow avalanches, snowstorms, tornadoes, tropical storms, tsunamis, wildfires, wind, and volcanoes. Because there is no single repository for hazard loss data in the United States, information was acquired from multiple sources. Data on the following categories were used:

Meteorological and related events with damage of at least $50,000, as reported in *Storm Data* for 1975–1994 (these data were input at below 5 percent error; single-loss estimates for multihazard events were ascribed to the first hazard listed; cost estimates were limited to broad categories of loss that could not be further refined in our analysis);[1] presidentially declared disasters from 1989 to 1994; insured losses from Property Claim Services from 1975 to 1994; losses from volcanic events that caused at least $50,000 in damage from 1975 to 1994; earthquakes from 1975 to 1994 that met one of the following criteria: moderate dam-

[1]Throughout this chapter all losses derived from Storm Data information have been converted to 1994 dollars, standardized on the basis of the Consumer Price Index. Unless noted otherwise, all figures from other sources are in actual dollars.

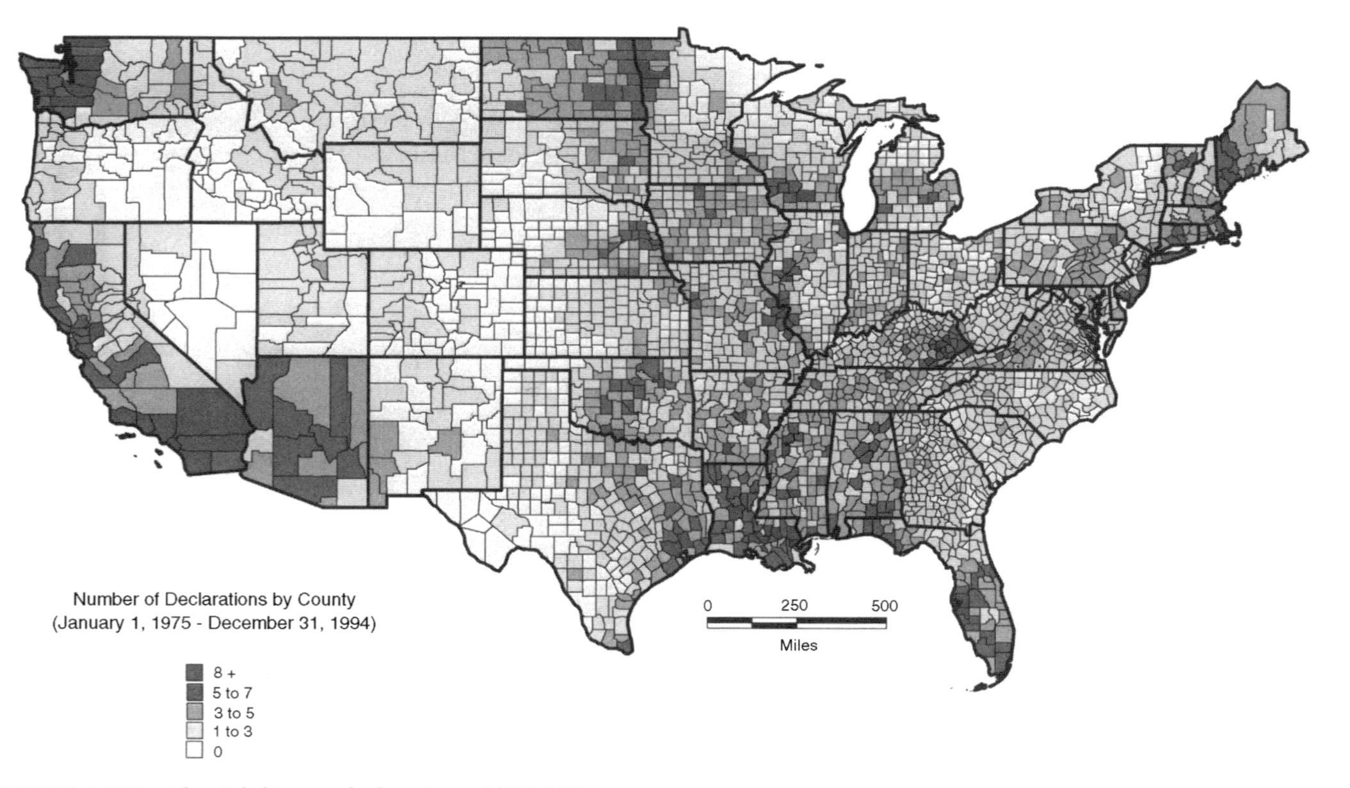

FIGURE 3.2 Presidential disaster declarations, 1975-1994.
Source: Federal Emergency Management Agency

age ($1 million or more), 10 or more deaths, magnitude 7.5 or greater on the Richter scale, or intensity 10 or greater on the Modified Mercalli scale; and losses and cost estimates from other official and academic sources, as appropriate.

There are a number of limitations to these data. The *Storm Data* database was the most comprehensive but also the most problematic. *Storm Data* is a monthly publication of the National Climate Data Center of the National Weather Service and is a comprehensive data source for climatological hazards. Each entry in this database accounts for one event, but one entry usually consists of multiple hazards. For example, if a thunderstorm produced wind, hail, and lightning, these were all attributed to the one hazard that most likely caused the losses, making it impossible to tell which hazard contributed how much to the total damage, deaths, or injuries. A single entry also could cover multiple counties. In addition, damage is reported in *Storm Data* as a range rather than an exact number, preventing a confident calculation of an overall sum or mean for a given period.

Data problems were not limited to the *Storm Data* database. Different agencies count losses in different ways. Some estimates are based on economic models, others on observers' reports. Some include estimates of indirect losses such as lost business income while others do not; still others use different combinations of these.

Climatological Hazards

Climatological hazards include those events that originate in the earth's atmosphere and precipitation that can cause harm to people and their property.

Droughts and Dust Storms

Droughts are caused by a natural reduction in precipitation over an extended period of time. They can occur in any climate and during any season; thus, the entire United States is at risk. Dust storms are usually caused by winds blowing over dry, loosely consolidated sediment not anchored by vegetation. Dust storms are more likely and more severe during droughts; they can further erode soils whose fertility has already been diminished by drought.

Storm Data reported no deaths or injuries from drought for the period 1975 to 1994, but 26 deaths and 151 injuries were reported as being due to dust storms. These data are difficult to interpret because other evidence suggests that human mortality and morbidity impacts are higher. For example, Trenberth and Guillemot (1996) report that 5,000 to 10,000 lives were lost in the 1988 drought, and a November 1991 dust storm in Fresno, California, killed 17 and injured 135.

The cost of property losses from droughts was $600 million to $6 billion; costs for dust storms were $2 million to $22 million. Drought destroyed $7.6 billion to $76 billion in crops in the United States in the 1975–1994 period, according to *Storm Data*. Other estimates of the costs of drought exist as well. For example, Riebsame et al. (1990) estimate that the 1987–1989 drought alone affected approximately 70 percent of the U.S. population at a cost of $39 billion. Losses reported in *Storm Data* to crops in the United States from drought for the period 1975 to 1994 peaked in 1976, 1988, and 1992. Missouri and Wisconsin were the two states that lost the most crops in 1976, while Iowa suffered the most during 1988. By 1989, Colorado, Iowa, Texas, and North Dakota also had suffered heavy losses. Average annual losses reported in *Storm Data* that were due to drought doubled (from $2.5 billion to $5 billion) during the 1975 to 1994 assessment period.

Losses from drought and dust storm hazards are likely to be both underestimated and inaccurate. Indirect losses from impacts such as farm foreclosures are not often counted, and direct losses, such as the destruction of annual crops, may be difficult to evaluate because of fluctuations in the commodities markets.

Extreme Cold

Extreme cold is defined as temperatures that remain cold enough to cause deaths, injuries, and damage to property and crops. According to *Storm Data*, 271 lives were lost due to extreme cold in the 20-year assessment period. The greatest number of deaths in any one year was 102 in 1983; the majority of these occurred in December in Arkansas and South Carolina. There was also an abnormally high number of reported injuries that year. A total of 522 injuries from extreme cold were reported for the 1975 to 1994 period. The worst year was 1994 with 184 injuries; of these, 129 occurred in Pennsylvania during mid-January.

The most costly year for property was 1994; losses were greatest in Ohio, Michigan, Pennsylvania, and Kentucky, and all occurred in January, before the spring growing season. The year 1977 was costly for both property and crops. Losses were greatest in February in North Carolina.

The drought of 1998 dried up this small lake outside Los Fresnos, a small town located at the extreme southern tip of Texas. Photograph by Brad Doherty.

Property losses from extreme cold showed a very slight upward trend over the 1975 to 1994 period, almost exclusively because of 1994 losses.

Extremely cold temperatures can be especially damaging to crops. The worst year reported in *Storm Data* for crop losses was 1989, when $700 million to $7 billion was lost; losses were heaviest in Florida and Texas. The next most costly year was 1990, when California was affected. Crop losses more than doubled over the 1975 to 1994 period.

There were four presidentially declared disasters due to freezes during the 1989 to 1994 period. Two declarations were granted for Florida and Texas in January 1990, one for California in February 1991, and one for Michigan in May 1994. The Federal Emergency Management Agency (FEMA) and *Storm Data* have different categories for hazards, making it impossible to accurately compare figures. FEMA reports declaration dates by state, while *Storm Data* reports events by state and county.

Floods

Floods were the most costly natural hazard in the United States in terms of deaths and dollar damage to property and crops over the 1975 to 1994 period. Floods can occur in all parts of the nation and at any time of the year but are more common in the spring when snowmelt combines with rain. The losses and costs from all kinds of floods are combined here because the reporting methods of different agencies do not always specify flood type.

Storm Data reports over 1,600 deaths from disastrous floods in the United States from 1975 to 1994. F. Richards (National Weather Service, personal communication, 1996) indicates that about 2,300 people lost their lives because of flooding in that period, and Pielke estimates deaths at 2,310 (1996a). FEMA (1991) reports an average of 78 deaths per year from 1925 to 1988, while almost 1,600 deaths were recorded from 1975 to 1985 by the National Weather Service (Riebsame et al., 1986). A peak in the death toll in 1976 was due to the July 31 Big Thompson Canyon, Colorado, flash flood, which killed 156 people. Another peak in 1985 was due to a flood in Puerto Rico that killed 180 people. Deaths were lowest in 1988 and 1989, which were drought years. Injuries declined slightly over the 1975 to 1994 period. They were highest in 1982 (1,078 in a two-day rain and snow event in California in January) and lowest in 1988 and 1989. The periods of fewest injuries coincide with those with the fewest deaths.

According to *Storm Data*, annual property losses from floods were $19.6 billion to $196 billion in the 1975 to 1994 period; Pielke (1996a) estimates losses for the same period at $67.5 billion (1992 dollars). Seven of those years had property damage exceeding $1 billion to $10 billion. Property losses rose over the 1975 to 1994 period. The most costly year was 1993, with property damage between $3.4 billion and $34 billion. Not surprisingly, 1988, the most costly year for droughts, was the least costly year for floods—only $31 million to $310 million.

An additional $8.1 billion to $81 billion in damage was done to crops between 1975 and 1994, and the trend indicates that losses are on the rise. More crops were destroyed by flooding during 1993, and 1988 had the fewest crop losses. Adding property and crop losses together, a total of some $27.7 billion to $277 billion was lost to floods during the 1975 to 1994 interval. Data from F. Richards (National Weather Service, personal communication, 1996) indicates flood losses of about $59 billion in 1992 dollars ($62 billion in 1994 dollars) for the same period.

From 1989 to 1994 FEMA paid out some $2.8 billion for disasters that included floods (excluding hurricanes and tropical storms). Often floods were part of multihazard events. (Note that FEMA and *Storm Data* use different collection procedures, so their data are not strictly comparable.) FEMA (1991) reports that from 1978 to 1990, approximately $2.5 billion was paid for flood losses insured under the National Flood Insurance Program (NFIP). Data from Property Claim Services (PCS) show that approximately $35 billion (actual dollars) and $41 billion were paid by insurance companies for losses caused at least in part by flooding. It appears that standardized dollar losses from floods are increasing, while injuries and deaths from disastrous floods are on the decline. However, the National Weather Service reports that the average annual number of flood deaths over the previous 25 years has increased steadily since 1968. Again, inconsistencies make it difficult to accurately assess losses.

Repetitive flood losses reported by NFIP statistics were assessed cumulatively from 1978 to 1994. The number of repetitive claims is highest in the Southeast, Midwest, mid-Atlantic states, and California. Louisiana had the most claims (104,000), followed by Texas (101,000). Alaska and Idaho had the fewest, under 10 per year. The percentage of repetitive claims is highest in North Dakota, the lower Mississippi Valley, and the mid-Atlantic states. In five states (Rhode Island, Virginia, New York, Louisiana, and Mississippi) over half of all the NFIP claims involved repetitive losses. Repetitive losses in Mississippi were about 65 percent.

Fog

Fog is essentially a cloud at the earth's surface that can reduce visibility to almost zero. Motor vehicle accidents are the primary cause of death and property losses because of fog. According to *Storm Data*, the first disastrous fog event in the 1975 to 1994 period occurred in 1984; only a total of 29 deaths were associated with disastrous fog for the 20-year period. However, based on data from the National Highway Traffic Safety Administration, Rosenfield (1996) reports 6,804 deaths in fog-related traffic accidents from 1982 to 1991 alone. The total number of deaths from all disasters from 1975 to 1994 as reported in *Storm Data* was 5,671. There are probably several factors that contribute to this glaring discrepancy. First, deaths and injuries from *Storm Data* were recorded only when losses exceeded $50,000. Second, although fog is obviously hazardous, it is not considered a storm and therefore may not always be reported in *Storm Data*. Fog may be a far more serious hazard than previously realized.

No doubt the injuries and property damage from fog found in *Storm*

Data are underestimated along with the deaths. For example, there were only 249 injuries attributed to fog from 1987 to 1994. The total fog-related property losses reported in *Storm Data* for the 1975 to 1994 period were $2.2 to $22 million. There were no reported crop losses from fog, which is probably accurate. Nor were there any presidentially declared fog disasters from 1989 to 1994. PCS has no record of claims filed for fog-related damage. It is likely that fog-related losses of all types are grossly underestimated.

Hail

Hail is ice that forms and grows in the updraft area of thunderstorms. It occurs in every state but mostly in the Great Plains. Because hail occurs only with thunderstorms, it falls most frequently during spring and summer, making crops and livestock particularly vulnerable. Hail can also damage property; cars are easily damaged by hail. Deaths from hail are extremely rare; injuries are more common.

According to *Storm Data*, there were 19 deaths from 1975 to 1994 in thunderstorms that contained hail. Deaths from hail are uncommon, so hail probably was not the cause of all 19 deaths. The Storm Prediction Center (1996) attributes only six deaths to hail during the 20-year period. Eagleman (1990) reports that only two people were killed by hail in the United States from 1930 to 1990. In 1930 a farmer was beaten to death by hail in Lubbock, Texas, and in 1979 softball-sized hail reportedly killed a baby in Fort Collins, Colorado. There were a total of 598 injuries from thunderstorms containing hail according to *Storm Data*. The Storm Prediction Center (1996) reports 294 hail-related injuries from 1975 to 1994.

According to the *Storm Data* information, total property losses from hail from 1975 to 1994 were $2.6 billion to $26 billion. The most costly year was 1990, with $600 million to $6 billion. Most of the 1990 damage was due to the July 11 hailstorm in Colorado, which passed directly over Denver's business district at lunchtime on a workday; 62 people were injured, and many automobiles were damaged. This remained the most costly hailstorm in U.S. history until May 5, 1995, when a hailstorm struck the Fort Worth, Texas, area, causing about $2 billion in property damage (Hill, 1996).

Total crop losses from hail for the 1975 to 1994 period were between $1.9 billion and $19 billion. The most costly year was 1982, with losses of $341 million to $3.42 billion. Two large storms accounted for most of the damage, one in California and one in New Mexico. Annual crop losses from hail have been estimated at $680 million (Eagleman, 1990), $130 million to $1.3 billion (*Storm Data*), and $667 million (Hillaker and Waite, 1985).

The two most costly hailstorms in U.S. history hit Denver, Colorado, and Fort Worth, Texas, with total damage exceeding $2.6 billion. Although this region is not as densely populated as others, it does contain prime ranch and farmlands. Consequently, crop and livestock losses from hail are high. The Midwest and South average more than 40 hail events per year, and their frequency has increased since 1975, peaking in 1992. Of the more than 64,000 hail events during the 1975 to 1994 period, only 2 percent resulted in damage greater than $50,000. Over the 20-year assessment period, crop losses from hail appeared to decline while property losses appeared to increase.

The total cost of presidentially declared disasters involving hail during the 1989 to 1994 period was about $9.2 million. However, rain, floods, wind, and tornadoes also contributed to this damage. According to PCS, insured losses involving hail totaled some $26.5 billion to $34 billion. This figure is misleading because hail losses were combined with other damaging elements of multihazard events such as wind, tornadoes, and sleet.

Heat

Heat is another weather-related natural hazard whose impacts on people and society are underestimated. According to *Storm Data* records, heat killed 674 people in the 1975 to 1994 period. Yet, for example, 295 heat-related deaths were reported in 1980 alone, just in Missouri (Rackers, 1996), none of which are recorded in *Storm Data*. Because heat does not always result in significant property or crop damage, the $50,000 threshold for counting incident-related deaths from *Storm Data* was not easily reached for this hazard.

Losses from heat as compiled from *Storm Data* reveal that the most lives were lost in 1980 when 570 people died; of these, 364 people perished in Alabama, Georgia, and Tennessee during July. Heat claimed another 136 casualties in September in Arkansas. The year 1986 was the worst for heat-related injuries, with 296 in Georgia at the end of July.

As might be expected, crop losses from heat were higher than property losses. Damage to crops totaled $1.7 billion to $17 billion over the 20-year period. Property losses add another $103 million to $1.03 billion. Most damage from heat occurred in 1980, when $90 million to $908 million of crops and $1.3 billion to $13 billion of properties were destroyed by heat. There were several years in which no losses were reported in *Storm Data*. In the past 140 years, 1988 was the fourth-warmest year globally (Christopherson, 1997), yet there were no disastrous crop losses from heat reported in *Storm Data*. There is no doubt that deaths, injuries, and costs from heat are underestimated.

Hurricanes and Tropical Storms

Hurricanes are predominantly East and Gulf Coast phenomena, yet their effects can be felt in other portions of the country. Hawaii and the West Coast also experience hurricanes. Before a storm attains hurricane status, it passes through the tropical storm stage where winds are between 65 and 119 kilometers per hour (hurricane winds are at least 120 kilometers per hour). Hurricanes that have been downgraded to tropical storms or depressions after moving over land are sometimes caught in a midlatitude cyclonic storm system. Then severe flooding can occur in noncoastal areas.

According to *Storm Data,* for the 1975 to 1994 period, hurricanes were the second most costly natural hazard in terms of property losses and the third most injurious. Because of advance warnings and emergency preparedness, hurricanes are only the seventh-leading cause of death due to natural hazards. During the 1975 to 1994 period, 173 people were killed by hurricanes and 23 people by tropical storms. Injuries from hurricanes totaled 4,525; 106 were due to tropical storms.

Over the 1975 to 1994 period the largest number of deaths from hurricanes in any one year was 43 from Hurricane Eloise, which struck Puerto Rico in September 1975. The next-highest number of deaths (25) occurred in 1992—the year of Hurricanes Andrew and Iniki. Andrew, which hit the east coast of Florida and the Gulf coast of Louisiana, caused 21 deaths. Iniki struck Kauai, Hawaii, in September 1992 and took three lives. In 1989, 17 people were killed in Puerto Rico, the Virgin Islands, and South Carolina by Hurricane Hugo. There was no change in average annual hurricane deaths over the 1975 to 1994 period.

The largest number of annual injuries from hurricanes occurred in 1983. Hurricane Alicia struck Texas in August, killing 13 people and injuring 1,800. In October 1985, 1,441 people sustained hurricane-related injuries when Hurricane Juan came ashore in Louisiana and Mississippi. There were five years with no reported injuries from hurricanes: 1978, 1981, 1987, 1990, and 1994. Only one injury per year was recorded in 1976, 1979, and 1993. There was no change in average annual injuries over the period.

The highest number of deaths from tropical storms was eight, from tropical storm Gordon in 1994. Gordon was also the cause of the second-highest number of tropical storm injuries: 43. The highest number of injuries was 50 in 1979, from a tropical storm that struck Virginia and Maryland, injuring 41 people.

Damage to property from hurricanes as reported in *Storm Data* over the 20-year period was between $11 billion and $111 billion; Pielke (1996b) estimates damage in this period at $60 billion. According to this information, 1989 (the year of Hurricane Hugo) was the most costly year for hurricanes, with $3.8 billion to $38 billion lost. The second most costly year was 1979, with $2.4 billion to $24 billion in losses from Hurricanes David and Frederic. The third most expensive year was 1992—the year of Andrew and Iniki, which together totaled $1.6 billion to $16 billion in property losses.

The National Oceanic and Atmospheric Administration (1996) estimates that Hurricane Andrew did $25 billion in damage while Hurricane Hugo $7 billion. Christopherson (1997) attributes $20 billion of damage to Andrew and $4 billion to Hugo. Ahrens (1993) states that Hugo cost $7 billion while Andrew cost over $20 billion. However, if crop damage from hurricanes is added to property damage, $2.7 billion to $27 billion was lost in 1992 to hurricanes Andrew and Iniki according to *Storm Data*. An analysis of *Storm Data* property losses in 1994 dollars inflicted by hurricanes shows an upward trend over the years 1975 to 1994.

Crop losses from hurricanes cost $4.6 billion to $46 billion, according to *Storm Data*, making hurricanes the third most damaging natural hazard over the 1975 to 1994 period. The most costly year for crop losses was 1989 (Hurricane Hugo), with losses between $1.3 billion to $13 billion. As with property damage, 1979 (Hurricanes David and Frederic) was the second most costly year, with $1.1 billion to $11 billion in crop losses. The year 1992 (Hurricanes Andrew and Iniki) was the third most expensive, with $1.06 billion to $10.6 billion in losses. The trend in annual crop losses is a very slight increase.

There was $592 million to $5.3 billion in damage to property and between $202 million and $2 billion in damage to crops from tropical storms. Property losses were greatest in 1979 and crop losses in 1989. FEMA named three tropical storms as presidentially declared disasters, the most costly two (Alberto and Gordon) in 1994. In total, FEMA spent some $420 million on tropical storm-related presidentially declared disasters. From 1989 to 1994 the U.S. government spent $4.2 billion on hurricane damage. Hurricane Andrew was the most costly, followed by Hugo.

Insured hurricane losses from PCS totaled approximately $24 billion. Andrew was the most costly hurricane, followed by Hugo. The year with the most hurricanes was 1985, when six hit the East and Gulf coasts. Estimated insured losses for these six events totaled $1.5 billion. The least expensive hurricane for the insurance industry was Babe in 1977, with insured losses estimated at $2 million.

Obviously, hurricanes and tropical storms are very costly. Fortunately, deaths and injuries appeared to decline or at least hold steady during the 1975 to 1995 period. But hurricane-related property, crop, and insured losses were on an upward trend. (Too few tropical storms occurred to determine a trend.)

Ice, Sleet, and Snow

Meteorologically, there is a difference between ice, sleet, and snow; however, for this study ice and sleet are combined into one category. Most losses from ice or snowstorms occur during winter, but they can occur anytime during the year in some parts of the country. Most of the snow in the United States occurs in the intermountain West, around the Great Lakes, and in New England. Most ice storms occur around the Great Lakes and in New England. The most severe snow/ice storm is called a blizzard—with winds of at least 35 mph and temperatures below 20°F (Eagleman, 1990).

According to *Storm Data*, snowstorms killed 729 people and ice storms killed 134 people from 1975 to 1994. By itself, snow is the fourth-largest killer. When combined, snow and ice are the third leading cause of deaths from storms. According to information compiled from *Storm Data*, the worst year for ice storm deaths was 1979, with 19. The most snowstorm deaths (92) occurred in 1978. There was a very slight decline in the number of deaths attributed to snow and in those attributed to ice storms over the assessment period.

There were 4,253 people injured by snowstorms and 1,039 people injured by ice storms in the 1975 to 1994 period, according to *Storm Data*. Together, snow and ice were the third-largest cause of injury during those years. The year with the most injuries from snowstorms was 1987, with 658. Almost one-half of the injuries occurred on October 4 in

A lull in the "Blizzard of '96" reveals snow-covered streets and buried cars in Silver Spring, Maryland. Photograph by Stephen St. John/NGS Image Collection.

New York. The next most injurious year was 1982, with 609 people injured, about half of those on January 10 in Alabama. Both snowstorm- and ice storm-related injuries increased over the 1975 to 1994 period.

Between $3.7 billion and $37 billion in property was lost because of snow and $1.1 billion to $11 billion because of ice and sleet. The most costly year for snowstorms was 1983, with property damage of $781 million and $7.8 billion. The most costly storm that year was in New York on February 6; it caused property damage of $500 million to $5 billion. Property damage from snowstorms increased slightly over the assessment period.

Between $1.1 billion and $11 billion worth of property was lost to ice and sleet storms. The worst year was 1994, followed by 1991 and 1978. These years do not correspond with costly snowstorm years. There were quite a number of ice storms in 1994, but the most severe occurred on February 10 and 11 in Kentucky, Mississippi, and Tennessee. Property losses from ice and sleet increased slightly over the 1975 to 1994 period.

Snow caused $215 million to $2.15 billion worth of crop damage, while ice was responsible for $660 million to $6.6 billion. Crop losses from snow apparently declined while crop losses from ice and sleet appeared to increase slightly.

Approximately $146 million was spent by FEMA on snow and ice-related presidentially declared disasters from 1989 to 1994. Snow and ice storms are associated with flooding, winter storms, freezes, and ice jams, so it is impossible to tell how much of the damage is due solely to snow and ice. Also, different reporting methods make it difficult to correlate these losses with those of *Storm Data*.

Lightning

Lightning is an electrical discharge that flows between positively and negatively charged surfaces. It is estimated that there are 100 lightning strikes on earth every second. Lightning is the most frequent cause of naturally ignited forest fires and, according to some, the deadliest of all natural hazards. Because lightning usually kills one person at a time, it tends to be underrated as a hazard.

The National Center for Atmospheric Research (1993) and the National Climatic Data Center (NCDC, 1996) confirm that lightning kills more people than any other kind of weather. R. Kithil (National Lightning Safety Institute, personal communication, 1996) reports that, from 1940 through 1981, 7,741 people were killed by lightning in the United States. The NCDC lists 1,444 deaths from lightning for the years 1975 to 1994. During this period, lightning was the second-leading cause of nondisaster deaths. The Great Lakes states, the Southeast, and Colorado had the largest number of casualties. As a single state, Florida had the most casualties for the 20-year period.

According to *Storm Data*, the total number of deaths from disastrous occurrences of lightning was 49 (vastly underestimated because *Storm Data* deaths were not counted unless they occurred along with monetary damages of $50,000 or more, and most lightning deaths are not associated with property or crop damage). Most of the reported deaths and injuries from lightning occured in 1987. There were no reported deaths from disastrous lightning strikes in 1986 and 1988, which documents how compilation of the *Storm Data* information underestimates losses from lightning when it is compared to the NCDC data.

Some 440 injuries are found in *Storm Data* that are attributed to disastrous lightning over the 1975 to 1994 period. Data from the NCDC indicate that from 1975 to 1994 lightning injured 6,158 people (again, the discrepancy in the figures is probably due to the lower cutoff point used in the collection of *Storm Data* information).

According to the NCDC, the total number of lightning events per year over the 1975 to 1994 period remained unchanged. The trend in deaths was slightly downward, according to both *Storm Data* and NCDC data. Lightning-caused injuries likewise had a downward trend.

According to *Storm Data*, property damage from lightning was between $0.41 billion and $4.1 billion. The most costly year for property damage was 1989, with about $150 million to $1.5 billion lost. A series of storms in California in 1987 appear to have caused the most damage. Apparently, property losses from disastrous lightning events remained stable from 1975 to 1994.

R. Kithil (National Lightning Safety Institute, personal communication, 1996) states that fires ignited by lightning cost an average of $137.8 million a year. If Kithil's figures are multiplied over the 20-year 1974 to 1995 period, the cost of lightning damage for that period would be $2.8 billion. Another estimate suggests that losses from lightning amount to $200 million per year (Riebsame et al., 1986). This figure is also congruent with the information from *Storm Data* and R. Kithil. Lightning-induced crop losses cost $20 million to $200 million. The most costly year was 1989, with $14.5 million to $145 million in losses. A very slight upward trend in crop losses from lightning was evidenced from 1975 to 1994.

R. Kithil (National Lightning Safety Institute, personal communication, 1996) estimates that insurance companies pay an average of $1 billion annually on homeowners' claims for lightning damage—an average of one claim for every 57 lightning strikes in the United States. Neither PCS nor FEMA reported paying out monies for lightning damage. Clearly, this is a very large discrepancy between datasets. The figures supplied by Kithil would increase losses from lightning by $20 billion, suggesting that lightning was one of the most costly natural hazards over the 1975 to 1994 period.

Snow Avalanche

A snow avalanche is a type of slope failure that can occur whenever snow is deposited on slopes steeper than 20 or 30 degrees (National Research Council, 1990). According to FEMA (1991), there are some 10,000 reported snow avalanches in the nation annually, but many others go unreported because of their remoteness. No deaths or injuries from disastrous avalanches in the United States were reported in *Storm Data* during the 20-year assessment period, but FEMA (1991) reports that, on average, 140 people are caught in nondisastrous snow avalanches annually and that 17 die.

No comprehensive record or even estimate exists of the economic costs of snow avalanches in the United States. Fragmented estimates include that of the National Research Council (1990), which indicates that approximately $330,000 is spent by the Washington State Department of Transportation on snow avalanches annually, while Alaskan avalanches from 1977 to 1986 cost $11.4 million. Snow avalanche rescue operations in Colorado on one day in 1987 cost the state $75,000; and a March 1981 avalanche in California caused $1.5 million in damage, killed seven people, and resulted in lawsuits totaling $14 million. Property damage attributed to disastrous avalanches in *Storm Data* for the 1975 to 1994 period was $100,000 to $1 million. The true cost of snow avalanches is not known, but losses doubtless will increase as development and recreation increase in mountainous areas.

Tornadoes

Tornadoes are the most violent storms on earth, and the United States has more tornadoes than any other country. These storms can occur any time of the year but are most common in the spring and summer in the Midwest, Southeast, and Southwest. The frequency of tornadoes remained constant during the 1975 to 1994 period, although there were slight increases during the last five years of the assessment period. The year 1992 holds the record for the most U.S. tornadoes, with 1,297 reported. According to *Storm Data*, tornadoes injure more people than any other natural hazard, are the second-leading hazard-related cause of death, and are the third leading cause of property damage.

According to *Storm Data*, total deaths from disastrous tornadoes for the 1975 to 1994 period were 1,090; the Storm Prediction Center (1996) reports 1,104. The year with the most deaths was 1984, when 130 people were killed. Of these, 58 were killed on March 28 in the mountains of North Carolina, South Carolina, and Georgia. The second most deadly year was 1994, when 83 people were killed on March 27 in Alabama, South Carolina, North Carolina, and Georgia.

Tornadoes are the number one cause of injuries from natural hazards. *Storm Data* indicates that 23,507 people were injured by disastrous tornadoes from 1975 to 1994; the Storm Prediction Center (1996) reports 25,006. Most of the 2,961 injuries in 1979 (the most injurious year) occurred on May 10 in Wichita Falls, Texas.

According to *Storm Data*, property damaged by tornadoes cost between $5.8 and $58 billion, making them the third-leading cause of property losses. Crop losses were lower, at a cost of $0.21 and $2.1 billion. The worst year for property damage was 1993, with from $0.66

billion to $6.6 billion in damage. Much of this was caused by tornadoes in Florida and Georgia on March 12 and 13. The costliest year for crops was 1979, when $38 million to $380 million worth of crops were destroyed.

Approximately $1 billion was spent by FEMA for presidentially declared disasters involving tornadoes from 1989 to 1994. Tornado deaths increased slightly over the 20-year period, but injuries appear to have decreased slightly. Crop losses also appeared to have decreased, while property losses were almost steady for the period.

Wildfires

Wildfires are uncontrolled burning in grasslands, brush, or woodlands. Lightning is the prime source of fires not ignited by people. All areas of the nation have thunderstorms, and fire danger can be high in a dry area that has available fuel. This is particularly true of the forested semiarid western United States, but humid areas can be at risk if there has been a drought.

Storm Data records wildfire losses only if they were a direct result of meteorological phenomena; thus, loss estimates are low compared to other sources. According to *Storm Data*, there were 10 deaths and 182 injuries as a result of disastrous wildfires from 1975 to 1994. Property damage was between $13.7 million and $137 million, with 1994 being the most costly year. During 1994 there were several unrelated fires in the western United States, but a fire in New Mexico on July 10 accounted for most of the year's damage. Crop losses totaled $61.8 million to $618 million. The most costly year for crops was 1992, when a wildfire in Idaho caused between $50 million and $500 million in damage.

According to Crosby (1993), seven of the nation's 14 worst wildfires since 1910 occurred during the assessment period: the 1977 Sycamore, California, fire that burned 805 acres and 234 homes; the 1980 Panorama, California, fire, which destroyed 23,600 acres and 325 homes; the Siege of 1987, California, fire which burned 640,000 acres; the 1988 Greater Yellowstone, Wyoming, fire, which destroyed 1.5 million acres; the 1988 Canyon Creek, Montana, fire, which destroyed 250,000 acres; the 1990 Painted Cave, California, fire that destroyed 641 homes and 4,900 acres; and the 1991 Oakland Hills, California, fire, which claimed 25 lives and burned 3,403 homes.

Other information also suggests that fire losses reported in *Storm Data* are underestimated. For example, a 1984 report by the U.S. Depart-

A stream of firefighters marches to the front lines of one of the 1998 Yellowstone fires. Thousands of firefighters battled the 387,000-acre fire. Photograph by Bob Zellar, courtesy of the *Billings Gazette.*

ment of Agriculture (1990) states that 2.82 million acres of protected forests were destroyed by wildfire in 1982 alone. These included commercial and noncommercial forests and nonforested watersheds. FEMA (1991) reports that 1988 was the worse single season for wildfires in 60 years. Federal expenditures alone were about $5.38 million. In all FEMA spent some $435 million on wildfires that were declared disasters by the President during the assessment period.

Insurance companies paid out approximately $3.1 billion in actual dollars and $3.7 billion for fires. The largest wildfire during the period was the 1991 Oakland Hills, California, fire, which accounted for almost half of all losses. If the Oakland Hills fire is excluded, no apparent trend in losses from fire is detected over the 1975–1994 period.

Wind

The entire United States is at risk from damaging winds. Winds are always part of severe storms such as hurricanes, tornadoes, and blizzards but do not have to accompany a storm to be disastrous. Down-slope windstorms, straight-line winds, and microbursts can all cause death, injury, and property and crop damage. This compilation includes all winds that either caused or were associated with natural disasters, regardless of category.

A total of 649 deaths and 6,670 injuries from disastrous winds were listed in *Storm Data* for the 1975 to 1994 period. Wind is the fourth-leading cause of death and the second-leading cause of injury. Deaths from winds were highest in 1975: out of 103 deaths, 31 occurred on November 10 in Michigan. The second-highest death total was in 1983, when 98 people were killed. That year was also the worst for wind-related injuries, with 622.

Between $5.8 billion and $58 billion worth of property—in events that cost $50,000 or more—was lost to winds over the 1975 to 1994 period. Wind-induced crop damage cost another $1.5 billion to $15 billion. Winds were the fourth-leading cause of property damage. Winds damaged the most property in 1990, when losses totaled $0.7 billion to $7 billion. Most of this damaged occurred during a windstorm in Texas on February 1. The next most costly year was 1982, when $1 billion to $10 billion worth of property was damaged by wind. The year 1980 was the worst for crop losses—$0.22 to $2.2 billion worth was destroyed. The trends in both property and crop losses from wind over the assessment period are downward.

FEMA reported that wind was involved only in two presidentially declared disasters from 1989 to 1994. The total spent on wind that accompanied severe storms and rain was approximately $97 million. The first instance was in 1992 in California and the second in 1993 in Washington state. More than likely, wind damage is included under the "storm loss umbrellas" of hurricanes, winter storms, thunderstorms, and tornadoes. PCS reported wind damage in almost all insured losses that were caused by meteorological events; these data were not tabulated because they could not be separated.

Geophysical Hazards

Geophysical hazards are events that emanate from the earth's crust or below that can cause harm to people or their property.

Earthquakes

An earthquake is a sudden motion or trembling of the earth caused by the abrupt release of slowly accumulated strain by faulting or volcanic activity. Earthquakes are very common, and thousands occur annually in the United States. Fortunately, most are small and cause no damage. The two most costly earthquakes during the 1975 to 1994 assessment period were both in California: the 1989 quake in Loma Prieta, which measured 7.1 on the Richter scale, and the 6.4-magnitude 1994 Northridge quake. The Northridge quake was the most costly natural disaster in the nation's history.

The greatest number of deaths from earthquakes in any one year was 62 to 67 in the 1989 Loma Prieta earthquake, according to Bolt (1993). This was closely followed by the 1994 Northridge earthquake in which 61 people were killed. In 1980 a series of smaller earthquakes (including that preceding the eruption of Mount St. Helens) killed 38 people. And according to Bolt (1993), eight people were killed in the 1986 Whittier-Narrows, California, earthquake, although NOAA (1996) reports no deaths from earthquakes in 1986.

Total injuries from significant earthquakes during the 1975 to 1994 period were 14,147 (Earthquake Research Institute, 1996). Bolt (1993) reports that 3,757 people were injured in the Northridge quake and that 45 were injured in the 1983 Coalinga, California, earthquake. Fifty people sustained injuries in the 1980 earthquake in Livermore, California, and 94 in the El Centro, California, quake in 1986.

According to NOAA (1996), total costs for 1975 to 1994 earthquakes were some $38.6 billion. This estimate is probably low because many other sources report much higher losses from the Loma Prieta quake. The vast majority of costs were caused by the 1994 Northridge earthquake. NOAA data indicates an increase in earthquake losses in the United States over the 20-year assessment period. The upward trend is probably due to an increase in population in quake-prone areas and an increase in seismic stress along California's San Andreas fault system (see Chapter 4).

Landslides, Subsidence, and Expansive Soils

Under the classification system used by the International Union for Geological Sciences, "landslide" is a blanket term for the downslope movement of a mass of earth, rock, or debris and includes many types of slides, falls, topples, avalanches, and flows, including mudflows and debris flows (Cruden and Varnes, 1996). Most mudflows begin on barren slopes after heavy rains. Landslides can occur on vegetated slopes that outwardly appear stable and can also be triggered by an earthquake. Areas at risk from landslides and mudflows include seismically active mountainous regions, such as those in Alaska, Hawaii, and the West Coast. Other areas at risk are mountains where wildfires have destroyed vegetation, exposing barren ground to heavy rainfall. Landslides are also plentiful throughout the Appalachians (Federal Emergency Management Agency, 1991).

Storm Data greatly underestimates landslide losses and costs, attributing only three deaths, two injuries, and $124 million in damage to landslides from 1975 to 1994, with total annual property losses from landslides between $11 million and $110 million. But others report average annual deaths as 25 to 50 people and losses of $1 billion to $2 billion (Schuster and Fleming, 1986). For example, in 1982 thousands of landslides (mudflows) were triggered by 24.3 inches of rain in a 34-hour period in the San Francisco Bay area. Twenty-five people died as a result of landslides and $66 million in damage was reported (Federal Emergency Management Agency, 1991). The 1980 eruption of Mount St. Helens in Washington state triggered a landslide of about 2.8 billion cubic meters of earth, killing five to 10 people (Schuster and Fleming, 1986), but these deaths were included in the volcano's toll in *Storm Data*. And direct damage from landslides in central Utah in 1983 has been estimated at over $250 million (Anderson et al., 1984). These are only a few instances of the many landslides losses documented elsewhere but not included in *Storm Data*.

According to PCS, there were no insurance claims filed specifically for landslides or mudflows between 1975 and 1994. However, landslides and mudflows were part of multihazard events.

Although subsidence is rarely a catastrophic or fatal event, it causes annual losses estimated at $125 million in the United States (National Research Council, 1991). It can be caused by natural processes, such as the dissolving of limestone underground, an earthquake, or volcanic activity, or it can be the result of human actions such as withdrawal of subsurface fluids like oil and groundwater or by underground mining. Between the various natural and human-caused types of subsidence, most areas of the United States are at risk (Hays, 1981).

Expansive soils—those soils and soft rocks that tend to swell or shrink as a result of changes in moisture—tend to produce gradual rather than disastrous impacts. But the monetary cost of this hazard is high, estimated at as much as $7 billion annually in the United States (Hays, 1981). Expansive soils are more prevalent in the central and western United States and along the Gulf Coast than in the East, although localized instances can occur elsewhere.

Tsunamis

A tsunami is a wave generated by sharp and sudden vertical impulses to the ocean floor. Tsunamis are produced by submarine landslides, earthquakes, or volcanic eruptions. These waves can travel up to 500 mph in deep-ocean waters, where wave height is usually about a meter. As the wave trough approaches land, wave heights of 30 to 100 feet can result in inland flooding. Tsunamis occur as repeated waves, with the second or third often the largest.

The area at highest risk from tsunamis in the United States is Hawaii since it is surrounded by tsunami sources from around the Pacific Ocean. On average, the Hawaiian Islands receive one tsunami per year, with a strong one every seven years (Federal Emergency Management Agency, 1991). Alaska, the West Coast of the United States, and American Samoa also are at risk. Tsunamis occur less frequently in the Caribbean, but the large exposed population coupled with lack of preparedness there make Puerto Rico, the Virgin Islands, and the U.S. East Coast seriously vulnerable.

According to FEMA (1991), there was no major damage from a tsunami during the 1975–1994 assessment period. It should be noted, however, that there have been major, damaging, deadly tsunamis in recent history—one in 1946 in Hawaii; one in 1960 in Hawaii and along the West Coast; and another in 1964 in Alaska, the West Coast, and Hawaii. Tsunamis at least as large as those—and more destructive—can be expected in the future, and increased development along the West Coast has exposed more people and property to this hazard.

Volcanoes

A volcano is an eruption from the earth's interior that produces lava flows or violent explosions that issue rock, gases, and debris (Federal Emergency Management Agency, 1991). Explosive volcanoes contain high amounts of gases and are the most dangerous. There are 29 active volcanoes in the United States, with explosive types along the West Coast and Alaska and in the Yellowstone area and nonexplosive types in the Hawaiian Islands.

During the assessment period the eruption of Mount St. Helens caused 60 deaths. Damage as reported by the Federal Emergency Management Agency (1991) was $1.5 billion (in 1994 dollars). Other sources say the direct property losses reached $2 billion (in 1994 dollars), of which $27 million was insured. The indirect costs have not been measured. Wright and Pierson (1992) report that over 180 dwellings were destroyed in a 1990 eruption of Kilauea. And in Alaska the 1986 eruption of Augustine disrupted air traffic and deposited ash on Anchorage.

Technological Hazards

The monitoring of technological hazards has been subsumed under the auspices of federal regulatory agencies that track hazardous waste, monitor nuclear power plants, and provide remediation for polluted areas. The records on such events are short. For example, the Toxic Release Inventory—which monitors toxic releases from industrial sources—only came into existence in 1987 and the Superfund in 1980. Moreover, the United States does not keep records of the deaths, injuries, and dollar losses associated with technological hazards and disasters.

Data on hazardous materials spills from transportation sources (highway, rail, air, and water) cover the period 1975 to 1994. There is a noticeable shift in the number of reported incidents from 1980 to 1981 because of a change in reporting requirements to the regulatory agency. Spills smaller than 5 gallons are no longer required to be reported after 1990. As new safety measures were enacted, the numbers of injuries and deaths declined (Cutter and Ji, 1997). However, estimated damage has escalated since the mid-1970s. Hazardous materials spills are concentrated in the industrial Northeast, California, North Carolina, Tennessee, and Texas. Pennsylvania had the highest frequency of incidents, averaging more than 1,000 per year, over the 20-year assessment period.

When the Superfund program was created in 1980, there were already 8,000 candidate sites waiting to be listed so that they could begin the process of remediation, with an average of 2,500 added each year (Hird, 1994). The National Priority List (NPL) represents the "worst" sites in the country—sites that pose an immediate danger to the health and welfare of host communities. There are currently 1,270 sites on the NPL. While some sites have been cleaned up, others have been added to the list. New Jersey had the greatest number of NPL sites (107), about 8 percent of the nation's total. Over a third of all the sites are located in New Jersey, Pennsylvania, California, New York, and Michigan. Because of the difficulty of estimating damage and costs, money spent on remediation activities under the Superfund program were examined instead. Estimates suggest that since 1980 billions of dollars has been spent on NPL cleanups (Hird, 1994).

Toxic Release Inventory (TRI) sites are those industrial facilities releasing toxic substances into the air, water, or land. There are more than 22,000 TRI-reporting facilities nationwide. The majority are in the industrialized Northeast and Great Lakes states, Texas, California, and North Carolina. When the pounds of materials released are mapped, a different picture of the hazard emerges. Most of the TRI emissions are concentrated in Illinois, Ohio, Indiana, Michigan, in the Southeast (South Carolina and Georgia are the exceptions), and in Utah. Utah has a large magnesium facility in Tooele County that contributes more than 75 percent of the state's total emissions.

OTHER IMPACTS

Disasters, particularly catastrophic ones, can do more than impose deaths, injuries, and dollar losses. They can also redirect the character of social institutions, result in permanent new and costly regulations for future generations, alter ecosystems, and even disturb the stability of political regimes. Costs like these rarely, if ever, are counted as part of disaster impacts. This section describes some of the economic, social, institutional, and environmental impacts of disasters, illustrated with cases from the past two decades.

Economic and Social Impacts

In the United States the impacts of natural disasters began their exponential escalation with the 1989 Loma Prieta earthquake. And the United States is not the only nation coping with disasters on this scale.

Hurricane Andrew

Hurricane Andrew was the third most intense storm to hit the nation in the twentieth century. It struck Florida on August 24, 1992, at 5:00 a.m. Despite heavy rainfall and storm tides up to 17 feet, most of the damage was caused by winds up to 145 mph, with gusts over 175 mph.

In Dade County alone 86,000 people lost their jobs, 160,000 people became homeless, and a total of 126,000 houses and 9,000 mobile homes were destroyed. Andrew hit Louisiana on August 26, with winds of 120 mph. Some 3,300 homes were totally destroyed, and 18,000 homes and many small businesses were damaged. The losses from Hurricane Andrew totaled $25 billion by some estimates and $30 billion by others (Peacock et al., 1997). Researchers attribute the economic losses largely to the migration of people to the Sunbelt region beginning in the 1960s (see Chapter 4).

Midwest Floods

In the spring, summer, and fall of 1993 sustained precipitation throughout a large section of the upper Mississippi River basin caused widespread flooding in the Midwest. This led to the declaration of several states as major disaster areas by the President and caused $12 billion to $16 billion in damage. Over half of the losses were from agricultural damage to livestock, crops, levees, fields, farm buildings, and equipment, with other damage to residences, businesses, transportation, and public facilities. Many residences and businesses outside the floodplain were damaged by basement flooding and sewer backups.

Kobe Earthquake

On January 17, 1995, an earthquake rocked Kobe, Japan's sixth-largest city, at 5:46 a.m., while most residents were asleep. The destruction from the 20-second, 7.2-magnitude (Japanese scale) tremor shocked the world. Some 6,308 people were killed (not including associated suicides), 350,000 people became homeless, 43,177 people were injured, and 215,000 homes and other buildings were badly damaged or destroyed (Noji, 1997). Kobe, which had been considered safe from earthquakes, found itself with overcrowded shelters, broken water mains, over 350 fires, inadequate supplies, and jammed roadways. A large section of a central subway station caved in, the Kobe port was largely destroyed by liquefaction, and the Hanshin expressway collapsed in five places and rolled over, an image seen in photographs around the world. Railroad lines for high-speed trains were damaged in 36 places along a 56-mile stretch. The total burned-out area of Kobe was estimated at 1 million square meters. In addition, 1.3 million households lost water,

860,000 lost gas, 2.6 million lost electricity, and 300,000 telephone lines were cut.

The economic losses from this quake, the most devastating earthquake in Japan since the Tokyo quake in 1923, are estimated at $100 billion or more. The cost of reconstructing the city alone was set at $64.2 billion, thus exceeding the cost of any other natural disaster in history. The extensive lifeline and infrastructure damage will continue to disrupt business activities for a long time, thereby increasing losses further.

Institutional Impacts

The far-reaching institutional impacts of disasters can include changed government policies, shifts in institutional and organizational arrangements, innovations in regulations, and even changes in international relations. Two of the three examples which follow describe institutional impacts for technological rather than natural disasters.

Three Mile Island

In the early morning of March 28, 1979, one of two nuclear reactors at Three Mile Island (TMI) near Harrisburg, Pennsylvania, malfunctioned. The TMI accident, which lasted about a week, resulted in 140,000 residents of the Harrisburg-Middletown area leaving their homes temporarily. Although there were no fatalities, the accident evoked great national attention and debate about the safety of nuclear power. Several investigations were carried out, including one by a commission appointed by President Carter, one by the Electric Power Research Institute, and one by the Nuclear Regulatory Commission (NRC). Most reports cited the cause of the accident as operator, or human, error. The Presidential Commission's report included serious criticism of the NRC, and President Carter called for a reorganization of and a new chairperson for the agency (see Rogovin et al., 1980; Presidential Commission, 1979).

One notable institutional effect of the accident was the emergence of new national regulations for nuclear power plants. In particular, the TMI accident led to the drafting of one of the most stringent emergency regulations, the *Criteria for Preparation and Evaluation of Radiological Emergency Response Plans and Preparedness in Support of Nuclear Power Plants*, also known as Regulation 0654. It includes the most detailed emergency planning in the nation. The new emergency planning policies resulted not only in better emergency plans for existing plants but also in no new nuclear power plants being built in the nation because the cost of litigation over the new emergency plans was prohibitive. For example, the cost of litigating the emergency plan for the Shoreham reactor on Long Island, including interest payments on construction loans while litigation went on, increased the price of the plant from $2.5 billion to $6.5 billion. TMI also halted new construction of nuclear power plants in the United States and ultimately led to the demise of the commercial nuclear power program.

Midwest Floods

The federal government's role in mitigation expanded after the 1993 Midwest floods. Armed with a commitment to mitigation, FEMA, under the direction of James Lee Witt, began a "buyout and relocation" program for flood-prone communities throughout the United States. Although relocation was not a new technique, never before had it been promulgated on such a large scale. The program permanently removes structures from the floodplain to ensure that they will not be continuously flooded and repaired in future years. The vacant areas left by the riverfront are then converted into open space or public parks. This program demonstrates a federal shift to nonstructural mitigation and away from reliance on structural remedies, such as levees and reservoirs. The buyout program is seen by many as part of a more comprehensive and sound approach to floodplain management and wetlands restoration. The tremendous impact and losses of the 1993 Midwest floods were a catalyst for this significant change in federal policy and programs.

Chernobyl

The Chernobyl nuclear power station exploded on April 26, 1986, releasing a radioactive cloud across Russia, the Ukraine, much of Europe, and the world. The world's worst nuclear accident was responsible for 125,000 deaths, including many children, pregnant women, and rescue workers. Since the explosion, cancer rates in the area have increased dramatically, and the death rate in northern Ukraine has risen 16 percent. Doctors and scientists expect that most cancers and other diseases resulting from weakened immune systems will not surface until the second decade after exposure. In other words, the death toll from the accident will probably rise significantly from now until 2010 (United Nations Department of Humanitarian Affairs, 1995).

Three undamaged reactors at Chernobyl remain in operation, and some Ukrainian officials downplay the effects of the 1986 accident and future dangers. Under international pressure, the Ukrainian president agreed to close the reactors in the year 2000 but only with financial help from the industrialized nations. The most seriously affected regions, such as the Ukraine and Belarus, are receiving some help with medicines and medical equipment from international nongovernmental organizations. The international community responded in many ways to the disaster, including humanitarian aid; pressure to close the reactors; demands for more information about the dangers, especially for residents of Eastern Europe; and monitoring of exposed residents to learn how their bodies react to the radiation. The disaster emphasized to the world the extent to which all ecosystems are connected. An international commission on nuclear power was formed.

Another notable institutional ramification of Chernobyl was its role in the dismantling of the Soviet Union. Some commentators believe that the seeming inability of the central government to prevent the disaster, or to cope with the problems it caused for millions of people afterward, demonstrated to the Soviet people the "flaws of the system" (see Medvedev, 1990). The Chernobyl accident may also have been the catalyst for vitalization of the environmental movement in the Ukraine. Because of his writings on the Chernobyl disaster, the Ukrainian political leader Yuri Shcherbak became a leader in the Ukraine's new environmental movement and contributed to the formation and popularity of the Ukraine's Green World Party.

HAZARD LOSSES BY STATE

To assess in a rough way the relative hazardousness of different parts of the United States, states were ranked according to frequency of events, deaths, injuries, and damage over the 1975 to 1994 period. Three different procedures were used to develop the composite totals: raw numbers, standardization by area, and standardization by population. The rankings for each component (frequency, casualties, damage) were combined to yield an overall ranking for each hazard. Where dollar estimates were not available, damage was calculated using the number of events costing more than $50,000. Only the frequency indicator was used for some technological events because figures on deaths, injuries, and damage were not available. The individual rankings by hazard were summed to provide an overall ranking for each state. However simplistic, this approach provides some interesting information on the regionalization of hazard losses in the nation.

Based on the raw data, the southern states are the most hazard prone, not only in terms of the frequency of hazard events but in the amount of losses. Many of these events are high frequency but with modest damage (such as hail and severe storms), but cumulatively they overshadow the more dramatic but less frequent disasters. The industrialized states in the Northeast are also high-hazard areas. This is again a function of the prevalence of severe storms as well as from releases and spills from hazardous materials facilities. Texas ranks as the most hazardous state, followed by Florida, Georgia, and Ohio. The least hazardous, based on the raw count, are Vermont, Delaware, and Rhode Island.

A slightly different pattern emerges when state size is considered. The industrialized Northeast region still stands out in the high-hazard category, but the southern states are now limited to those in the Southeast. Ohio, South Carolina, and Pennsylvania are among the most hazardous states based on hazards per square mile, while Nevada, Alaska, and Montana are the least hazardous.

Each state was also standardized by population size to determine a relative measure of the potentially exposed population. Residents in the South, portions of the Great Plains, and Rocky Mountain states bear a disproportionate burden of hazardous events and losses on a per-capita basis. This is partly explained by the low population densities in many of these states. However, it also reflects the number of more common hazard events such as hail, severe storms, and wind. In this per-capita measure, Kansas is the most hazard-prone state, followed by Arkansas, Georgia, and South Carolina. The least hazard-prone are the Pacific Coast states.

Finally, and perhaps most importantly, a proportional measure was developed that permits an examination of the relative impact of hazards in each state. The percentage of a given hazard for a given state was calculated by taking the total number of specific hazardous events divided by the national total. This was also done for casualties by hazard and losses by hazard. The three indicators (reported as percentages) were summed across all hazards and an average was taken. Figure 3.3 maps the relative hazardousness of the nation using this "all-hazards" indicator. More than half of the states are in the low-hazard category. Regionally, this includes the western half of the country, upper Great Plains, New England, and the mid-Atlantic region. States that rank high in proportional damage and casualties are California, Texas, and Florida. The Southeast, lower Great Plains, and the Northeast states have more fre-

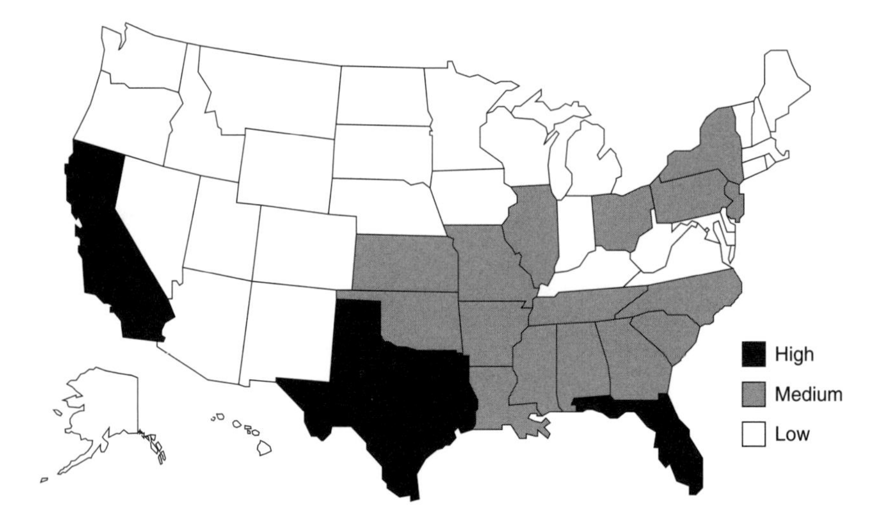

FIGURE 3.3 Rankings of hazardousness, by state, 1975 to 1994

quent natural hazard events, but they are less costly in terms of losses and casualties.

Figure 3.3 illustrates the difference between the more "ordinary" hazards like severe storms, tornadoes, and hail, which occur frequently but whose impact (based on losses and casualties) is more restricted. The extensive urbanization and high population density of Florida and California, coupled with their high risk potential, clearly explain their top rankings. This also illustrates the interactive character of hazards as discussed in Chapter 4. It takes only one or two events (an earthquake, major hurricane, major flood) to produce significant losses in the built environment in these places. Texas also ranks high because of the sheer number and diversity of hazardous events that occur there as well as the state's large although more dispersed pattern of urbanization.

ISSUES IN LOSS AND IMPACT MEASUREMENT

It should be clear from the loss data presented in this chapter that measuring the effects that disasters have on localities and the nation is fraught with uncertainty. Just gathering tallies and estimates that have already been made, presumably by experienced sources, is a formidable chore. Yet underlying even seemingly credible financial estimates is a range of complex issues about how a monetary value can be placed on what has been lost or diminished—or sometimes on what has been gained—as a result of a disaster.

Assigning Value to Losses

Most of the damage or loss figures quoted in this chapter and elsewhere are measures of direct expenditures or market valuations of goods, structures, facilities, and the like that have been lost or reduced in value because of a hazard. Dollar loss estimates tend to include only losses that have been valued as a result of market transactions. This excludes the very losses that are most poignant, such as lost memorabilia, destruction of historic monuments and cultural assets, environmental degradation, and the hidden cost of trauma. At present there is no satisfactory way of assigning value to these kinds of losses, although it certainly can be argued that an effort should be made to account for them in some, albeit imperfect, way.

Measuring Losses

Damage figures are only as good as the assessments done to arrive at them. Depreciated replacement value is the appropriate basis for assessing damage to assets, but most damage assessments do not account for it. If an asset is completely destroyed and replaced, "damage" equals the replacement cost. Accounting for depreciation can be simpler in theory than practice, however. Depreciation can be functional, aesthetic, economic, or physical. Older structures and equipment may have lost some functional and/or economic value and thus be more costly to maintain and operate. This added expense detracts from their market value. On the other hand, older residential buildings may sell for a premium, reflecting architectural methods and/or other special characteristics no longer commercially available. In addition, sometimes the replacement capital will embody technologies that will make future production more efficient, thus lessening the loss.

Another problem is that it is all too easy to double or even triple count some losses. For example, should the losses sustained by a bank that provided financing in a damage-stricken area be added to the property damage tally? Or is the bank's loss just its share of the total? A common mistake is to incorrectly add lost business income to lost wealth (Howe and Cochrane, 1993). A careless damage assessment might combine lost asset value (a stock measure) with lost income to the asset's owners (a flow measure). But income is derived from ownership, and an asset is properly valued in proportion to the income stream it generates. To count both would be incorrect. The same logic applies to stock prices, bank profitability, and insurer profitability. In each case a firm's worth is adjusted by the amount of direct loss sustained.

Direct and Indirect Damage

The word "damage" is used loosely in the hazards field (and often is used interchangeably with "loss"). Damage from hazards can be direct, indirect, market, nonmarket, or residual. Most loss estimates quoted here are for direct damage. The term "direct damage" means any loss that is attributable to the destruction of buildings, machinery, or public infrastructure. Employment losses traceable to the destruction of a workplace are considered a direct loss. "Indirect" loss (or gain) is anything other than direct loss. For example, factories relying on just-in-time inventory systems may be forced to close if a critical parts supplier cannot meet its obligations. Income lost as a result would be considered indirect damage.

Indirect loss was identified in the first assessment as a disaster-induced change in a region's trading pattern. At that time the concept was applied rather narrowly to include only the secondary ripple effects triggered by supply bottlenecks (i.e., the shortages that cause undamaged producers and service providers to shut down; Cochrane et al., 1974). Since then the concept has been refined, and the analytical tools for estimating such losses have advanced markedly (National Institute for Building Sciences, 1997; Rose et al., 1997). Yet despite such advancements, indirect loss remains a somewhat amorphous concept. In contrast to what was once believed, it does not readily yield simple rules of thumb (e.g., "indirect loss is equivalent in magnitude to direct loss"). In fact, it has since been learned that indirect loss can vary dramatically and under some circumstances may even be manifest in the form of regional benefits (negative losses). The concept of indirect loss can be more broadly applied than was first thought.

Economists now know that there is no simple linkage between direct and indirect real damages. The magnitude of indirect losses from a given disaster is highly dependent on the starting point—the preexisting conditions (how open the economy is to trade, excess capacity, the damage pattern) and, perhaps most important, how much outside assistance pours into the impacted region (insurance payouts and federal aid). It is also influenced by where the economic boundary is drawn—city, standard metropolitan statistical area, state, or nation.

Systematic efforts to collect data on indirect damage began only recently. The evidence analyzed to date points to a sizable regional economic expansion the year after an event (as with the Northridge earthquake, Hurricane Andrew, and Hurricane Hugo). It should be noted, however, that these results are preliminary and are for a limited time frame. Nonetheless, it appears that bottlenecks and associated indirect losses are less of a problem than once thought.

Interpreting Loss Measurements

Because there are so many subtleties inherent in using and interpreting damage statistics, it is no wonder that considerable confusion permeates the economic hazards literature. Statements about the economic impacts of hazards and disasters must be considered in the context of their use. If the numbers are reported to convince people that losses are too large and deserve to be ameliorated through government assistance, they would have to be compared with other indicators (annual wear and tear on the nation's capital stock, for example) to determine whether federal attention is economically justified. If they are used to compare events or measure progress over time, standardized dollars must be used. Comparisons with other meaningful measures (such as gross national product [GNP] or percentage of income) can be helpful. These comparisons have their own pitfalls, however. GNP, for instance, generally fails to identify losses of time, anxiety, loss or diminution in value of natural resources, wildlife habitat, and all of those "soft" economic losses from disasters, but it does count all reconstruction costs as gains. Looking at this "balance sheet" one must conclude that a disaster is a good thing, at least economically.

In another example it has often been pointed out lately that until the Loma Prieta earthquake in 1989 the insurance industry had never experienced a loss greater than $1 billion from a single event. Since then there have been 15 disasters exceeding $1 billion in insured losses. This is an impressive statistic when viewed in isolation. However, when normalized for population growth, wealth, inflation, and growth in the amount of surplus available to the insurance industry, such facts are not so amazing. For example, a $1 billion event in 1970, had one occurred, would have been the equivalent of a $15 billion event in 1992. Viewed in this context it is not so remarkable that an event that large had never before occurred. Furthermore, such loss potential had been recognized for some time. Damage statistics and simulation results published in the first assessment (White and Haas, 1975) indicated that hurricane damage could produce losses in excess of $50 billion (in 1997 dollars) annually.

Balancing short- and long-term measures of disaster impacts is another problematic area—one that is discussed more fully in Chapter 7.

There is considerable challenge in selecting the appropriate level of measurement—local, regional, national, or global—for a particular purpose. The national economy is affected by any costs incurred outside a stricken region and thus can negate local gains. This in turn can be offset somewhat by increases in outputs elsewhere. Direct damage and subsequent indirect loss are transmitted to other regions via altered trading patterns (imports and exports). If a region produces critical materials for which ready substitutes are unavailable, local indirect losses can spill over to affect the national economy. Hence, it is even possible for indirect loss to exceed that sustained by the region alone. So from the vantage point of the nation as a whole, indirect loss must exist. Can disasters serve to stimulate an economy and therefore be considered beneficial? A region may benefit, but the nation always suffers. Indirect gains are an illusion, a byproduct of focusing only on a stricken area.

FUTURE LOSSES

Catastrophic losses from disasters in the United States and the world are increasing. Globally, the annual number of disasters that cause 25 deaths or more tripled since the 1940s (Glickman and Golding, 1992). In the United States seven of the 10 most costly disasters in history, based on dollar losses, occurred between 1989 and 1994: Hurricane Andrew in 1992; the Oakland, California, wildfires in 1991; a winter storm in 1993; Hurricane Iniki in 1992; and the 1989 Loma Prieta and 1994 Northridge earthquakes (Federal Emergency Management Agency, 1995). Globally, the list would include the $120 billion earthquake in Kobe, Japan, in 1995.

The disasters of the past two decades, especially those since 1989, were undeniably destructive and immense. But it is clear that these events are not the "big ones." For example, it is estimated that a 7.0-magnitude earthquake in the Los Angeles area would result in up to $250 billion in economic loss, up to 5,000 deaths, and up to 15,000 serious injuries (Shah, 1995). Or consider that, although Hurricane Andrew was the costliest disaster in history at the time, officials note that it could have been much worse. As such storms go, Andrew was compact, fast moving, and relatively dry. If it had been slower and wetter or had its path been just 10 miles farther north (through downtown Miami), its devastation would have been far more widespread and profound.

While fatalities in developed countries have steadily decreased by 75 percent during the past 50 years (Mitchell, 1995), economic losses are rapidly increasing (Berke, 1995; Munich Reinsurance, 1995). New records for the costliest natural disaster are now set regularly. In 1989 Hurricane Hugo exacted losses of $6 billion; in the same year the Loma Prieta earthquake cost $10 billion. In 1992 Hurricane Andrew cost $20 billion. Losses from the 1994 Northridge earthquake were $25 billion, and the 1995 Kobe, Japan, earthquake cost more than $100 billion. Currently, direct losses from natural disasters in developed nations represent only a small fraction of GNP. For example, Hurricane Andrew and the Northridge earthquake each represented one half of 1 percent of the GNP of the United States. But if the trend of increasing losses continues, individual disasters will represent more significant fractions of the GNP of affected countries and affect global economies, largely because of increased social complexity and the interconnectedness of global societies.

An often quoted and controversial scenario of future losses is that of the effects on world financial markets of a repeat of the 1923 Tokyo earthquake. Current loss estimates approach $1 trillion or approximately 40 percent of the current Japanese GNP (Shah, 1995; Wiggins, 1996). In 1989 the Tokai Bank estimated that the withdrawal of foreign investment needed to rebuild Tokyo after such an event would amount to 1 to 3 percent of the annual GNP of the United States and up to 14 percent of Latin American and European GNPs. It is not clear whether a future great earthquake in Tokyo would destabilize world capital markets.

The magnitudes of loss estimates are greater when considered at lower levels of aggregation. For example, while direct losses in the Northridge earthquake accounted for only 0.5 percent of the GNP, they represented approximately 3 percent of California's economy (based on 1992 gross state product [GSP]). Similarly, losses from Hurricane Andrew represented approximately 7 percent of Florida's economy (based on 1992 GSP).

Would a series of catastrophic natural disasters destabilize the municipal bond market? Insurance industry economists have stated that the sale of some $30 billion worth of paper on the municipal bond market at one time would have two immediate effects. First, the industry would never get 100 cents on the dollar. Such a move would severely strain the municipal bond market, probably closing it altogether. This would mean that a city just devastated by a quake would have nowhere to go to raise money. It may be correct that the industry will not get 100 cents on the dollar, but only marginal price reductions are likely to occur. Municipal bond yields rise and fall with U.S. Treasury Bill rates and the anticipated tax advantages that municipals offer. Destabilizing events such as New York's fiscal crisis from June 1975 to December 1976 had no impact on this market. In the unlikely event that dumping produced capital losses for the insurance industry, these would be matched by capital gains elsewhere. Such effects cancel each other when viewed from a national perspective. Furthermore, price reductions (yield increases) represent a normal market response to a shock of this nature. Interest rates need to rise to attract the domestic and foreign savings required for reconstruction. Lastly, if the industry believes that such losses are a problem, companies can diversify their portfolios to include other liquid instruments.

In summary, capital markets are probably too vast to be disturbed by natural disasters. The municipal bond market, which had been thought to be sensitive to sudden liquidations, seems to be moved by interest earned on alternative investments and current and proposed tax laws. Most movements in stock prices after a disruptive event simply mirror the real losses that corporations sustain. As such, stock price movements are a measure of direct loss, not indirect loss (Cochrane et al., 1992).

DATA NEEDS

Although this chapter has been filled with numbers about losses and costs from disasters and hazards, the truth is that losses to the United States and its people from natural and technological hazards are not really known with much certainty. And as discussed in the previous section, it is not necessarily clear what the loss figures mean either. The nation must anticipate future disasters, aggressively adopt mitigation strategies, and then evaluate their effectiveness. A large amount of accurate standardized information will be needed to do that.

What is needed is a comprehensive database that contains information about (1) current levels of vulnerability to natural hazards on national and local scales, (2) compilations of past losses, and (3) the costs of pre-event mitigation activities. Existing vulnerabilities inherent in both the constructed and the social environments must be estimated.

A national building inventory, in a geographic information system-type format, listing such parameters as location, age, size, current value, and type of construction, combined with geophysical estimates of the likelihood and severity of natural hazards, would enable assessments of national and local risk. These assessments would help target areas for mitigation. Previous loss records only indicate in a general way the overall scale and scope of the problem. Monetary losses have not been systematically assessed, nor have the economic ramifications of a disrupted social structure been compiled. The next data generation requires better delineation of the types and extent of losses in specialized categories. A national loss inventory would document losses from past and current natural disasters, thereby establishing a baseline for comparison with future losses. Data on the type of loss, location, specific cause of the loss, and actual dollar amounts need to be compiled in a uniform fashion for across-hazards comparisons.

A long-term perspective is necessary if the mitigation practices of the 1990s are not to contribute to future hazards. Past mitigation efforts should be cataloged as a baseline, along with their effects on loss reduction. Although FEMA has begun to document the procedures and costs associated with successful mitigation projects, there is a woefully small amount of that kind of essential information available today. Such a database would allow comparisons of alternative plans in the context of past losses, future risk, and dollars needed to be spent today to trade off against losses tomorrow. Furthermore, there should be public access to the data so that everyone has the opportunity to make more informed decisions. The Internet and World Wide Web provide the ideal repository for a national database.

Data should be collected by a standardized system, taking into account warnings by economists about double-counting, standardization, comparisons with gross national product, distinctions between regional and national impacts, and other factors.

CONCLUSION

The United States has succeeded in saving lives and reducing injuries imposed by most natural disasters such as hurricanes over the past two decades. Lightning and tornadoes are unfortunate exceptions: these two natural hazards continue to take lives and impose injuries just as they did two decades ago. Nor have proportional decreases been achieved in the death and injury tolls from floods—the nation's largest natural hazard. It is dismaying that two hazards largely ignored by most of the hazards community over the past 20 years—fog and heat—have negated the gains in lives saved and injuries avoided on other hazards fronts.

Clearly, the dollar losses associated with natural hazards and disasters in the United States rose over the past two decades. To some extent this is not surprising because of the nation's dramatic increases in capital stock—there is more to lose now than there was 20 years ago. But the nature of hazards also is changing and is growing more complex. The ways in which this is happening, and the ways in which it manifests itself in increasing losses, are discussed in Chapter 4.

An integrated accessible database of losses and vulnerabilities could create a feedback loop between communities, researchers, regulatory agencies, emergency managers, the insurance industry, and others to improve overall effectiveness in coping with hazards and disasters. It will be essential if the nation is to steer an informed course toward sustainable hazards mitigation into the next century.

CHAPTER FOUR

The Interactive Structure of Hazard

LOSSES FROM HAZARDS and disasters in the United States in the next millennium will be determined, as in the past, by a large number of variable factors. These factors can be grouped into three broad categories: the natural environment, the social world, and the human-made constructed environment. Logically, then, hazards mitigation in the next millennium should be based on a triad structure of hazard that takes into account (1) the influences exerted by the physical environment, (2) how human systems create and redistribute hazards, and (3) hazards that result from the nature of the constructed environment.

This recommendation is not a new one. White and Haas (1975) made a similar argument in the nation's first assessment. They pointed out that the hazards field, until that time, had ignored "people factors," meaning the social, economic, and political dimensions of hazards. They also noted that the nation had concentrated too heavily on technological solutions when what was needed was "technology in balance" (p. 2). They suggested that one way to reduce vulnerability was to use programs that provide equitable distributions of costs and benefits of disaster recovery among various economic and social groups. They pointed out the role played by individual material wealth in determining unequal vulnerability to disaster losses. This banner was taken up to some degree by the social vulnerability school of hazards research (described later in the section "The Human System"), particularly for work outside the United States, but the mainstream of hazards research and management in this country is only now coming to a realization of the important role that human-constructed social realities play in creating disasters.

This chapter examines those three systems—natural, human, and constructed—some of their components, and how their actions and interactions—both inside and within each system—produce a constantly changing level of hazardousness for the United States and the world. It concludes with a call for a national hazards risk assessment, including taking stock of all three systems discussed in this chapter from the largest global processes down to the individual.

It should be noted that the terms "risk," "exposure," and "vulnerability" as used in this chapter have specific meanings, adopted from Mitchell (1990). Risk means the probability of an event or condition occurring. Exposure is the measure of people, property, or other interests that would be subject to a given risk. Vulnerability is the measure of the capacity to weather, resist, or recover from the impacts of a hazard in the long term as well as the short term. Hazard can be loosely thought of as the product of risk, vulnerability, exposure, and the capacity of humans to respond to extreme events.

SYSTEMS

Hazards researchers and practitioners would do well to take a more systems-based approach to understanding the complex interactions between the natural environment and human perceptions, actions (including what people build and where it is located), and organizations. Systems theory, first proposed in the 1940s, is a broad interdisciplinary domain. It foreshadowed and is related to theories of complexity, chaos, cybernetics, and complex adaptive systems. The basic idea is that all complex entities—be they biological, social, ecological, or other—are composed of numerous diverse elements linked by strong interactions. But a system is greater than, and different from, the sum of its parts. This differs from the traditional analytical scientific model, which follows the laws of additivity of elementary properties (the whole is equal to the sum of its parts).

Because additivity does not apply to them, complex systems must be studied differently, holistically. A systemic approach to problems focuses on interactions among the elements of a system and on the effects of its interactions; it examines a variety of factors at one time; it integrates time, feedback, and uncertainty. Systems theory recognizes multiple and interrelated causal factors, emphasizes process, and is particularly interested in the transitional points at which a system—be it a hurricane, a communications network, or a society—is open to potential changes. Traditional models, on the other hand, are typically linear and assume only one causal factor. They tend to overemphasize stability and end up labeling change and process as negative. Using the systems approach allows a wider variety of factors and interactions to be taken into account.

Viewed from this perspective, disaster losses are the result of interaction among three systems and their many subsystems:

- the earth's physical systems (the atmosphere, biosphere, cryosphere, hydrosphere, and lithosphere);
- human systems (e.g., population, culture, technology, social class, economics, and politics); and
- the constructed system (e.g., buildings, roads, bridges, public infrastructure, housing).

Many different interactions are possible both within and between earth systems, social systems, and constructed systems. All of the systems and subsystems are dynamic in and of themselves. In addition, there are more or less constant interactions between and among two or more subsystems and systems. Finally, almost all human and constructed systems—and some physical ones—are becoming more complex with each passing year. This complexity is what makes national and international disaster problems so difficult to solve. And the increase in the size and complexity of the various systems is what causes increasing susceptibility to disaster losses. Changes in size and characteristics of the population and changes in the constructed environment interact with changing physical systems to generate future exposure and determine losses in future disasters. The world is becoming increasingly complex and interconnected, helping to make disaster losses greater.

One brief example hints at the range of complexities involved in interactions among earth, social, and constructed systems. In 1971 changing international politics increased the economic demands of the Cold War. In earthquake-prone Armenia, steel originally intended to reinforce family housing was used to make weapons instead. This left the structures in the city of Leninakan vulnerable to even moderate ground shaking. At the same time, a high rate of population growth due in part to religious beliefs increased the density of people living in buildings constructed since 1971. The loss of 25,000 lives in the 1988 Spitak quake thus was the result of interactions among construction practices, international political conditions (the Cold War), culture (religious beliefs), and the lack of economic resources (funds to relieve housing density).

The Earth System

The earth system comprises the natural and physical elements of the earth and its atmosphere that exist and change over time. Extremes in the system pose natural hazards and can become disasters when they affect humans.

The atmosphere is the gaseous subsystem of the earth system. Most "weather" occurs in its lower 11 to 12 kilometers and can include hurricanes, tornadoes, blizzards, hail, and lightning. The biosphere is composed of all living matter (excluding humans for this discussion). The cryosphere is that part of the planet covered by ice and/or snow. It is important in the earth's radiation balance and can generate social losses through snow avalanches, permafrost, and icebergs. The hydrosphere is the liquid portion of the earth system that includes both fresh- and saltwater bodies. Losses from the hydrosphere are usually the result of flooding. The lithosphere is the solid portion of the earth system and is covered by air, water, and/or ice. Losses from solid earth processes come from earthquakes, volcanoes, landslides, erosion, and deposition.

Two subsystems of the earth system are described here as illustrations of the kinds and numbers of interactions that take place within physical systems that have eventual implications for natural disasters. The first is a global subsystem—climate. The second is far more localized—seismicity in California.

The Case of Climate

Global climate consists of a number of closely coupled subsystems, including the atmosphere, oceans, terrestrial biosphere, and cryosphere. The influx of solar energy into the atmosphere drives a range of climate processes, including evaporation, condensation, cloud formation, precipitation, and runoff; atmospheric and oceanic motion and circulation; and the formation and melting of snow and ice. These processes involve fluxes of energy and matter between the climate subsystems on many different spatial and temporal scales. Finally, to avoid overheating the planet, the earth releases radiation from its surface and atmosphere. While insulating properties of the atmosphere help slow the heat's escape, ultimately heat loss to space must equal the energy absorbed from the sun if the global climate system is to remain in equilibrium.

Climate is a function of many nonatmospheric factors as well, such as topography, ice sheets, and ocean circulation. This synthesis includes not only averages of weather but also estimates of its variability and the probabilities of extreme events within a region and is usually characterized by statistics such as means, standard deviations, variance, and return periods for extreme events (Maunder, 1986).

The climate of a particular region is not a constant. The mean annual temperature of the northern hemisphere warmed several tenths of a degree during the 1930s and 1940s, then stabilized or cooled slightly for several decades, followed by a warming since the mid-1970s. Much of North America has experienced an increase in average precipitation, particularly in winter over the past few decades. On longer time scales, changes in mean global temperatures of 4 to 5°C have occurred at intervals of about 100,000 years, with shorter and smaller fluctuations in between. Calculations for climate statistics are commonly based on the three most recent full decades, but prudent users of such data recognize the influence of longer-term natural variability.

Some of the global factors that can affect climate are well defined and the direction in which they push or "force" the global climate system can be calculated and modeled quite accurately. An increase in solar energy entering or being stored in the atmosphere, for example, will push the climate system toward warmer average temperatures, while the addition of highly reflective volcanic dust into the atmosphere would tend to cause cooling by reflecting sunlight back to space.

Somewhat more difficult to estimate are the forcing effects of changes in regional concentrations of atmospheric constituents that stay in the air a relatively short time and whose radiative properties are more variable. For example, dust and other solid particles (which tend to remain in the atmosphere only for days or weeks) can both reflect and absorb sunlight, depending on their size and other properties. Localized changes in concentrations of ozone in the lower atmosphere (owing to urban smog) can affect heat flow budgets by enhancing the local greenhouse effect, but such trends are difficult to measure and the net global effects cannot be estimated accurately.

Climate Change. In a technical sense, climate change refers to a significant change in climate statistics for a given area, regardless of cause. Such changes would also include the frequency and severity of extreme events and hence the hazards of weather. However, changes in climate statistics calculated for a 30-year period as identified by a climatologist, for example, may be considered simple climate variability by a glaciologist studying long-term records.

The climate system is complex and nonlinear. Changes in the primary flow of energy into and out of the system caused by radiative forcing factors (like those discussed above) result in second- and third-order effects on many of the processes, which in turn can affect the changes that caused them. Such "feedbacks" can amplify or attenuate the original effect. Particularly important in the short term—days to weeks—are feedbacks linked to such variables as water vapor content in the atmosphere, cloud cover, soil moisture, and snow and ice cover. On longer time scales—years to decades and beyond—changes in vegetation, ocean fluxes and circulation, continental ice mass and permafrost, or other variables that change slowly also have strong feedbacks. These feedbacks can vary substantially from one region to the next and result in a nonuniform distribution of climate changes that is difficult to predict at the less-than-global level. Hence, while models used to simulate global climate change may operate quite accurately for the global scale, confidence in their predictions decreases rapidly with scale and with respect to second- and third-order impacts.

The response of the climate system to changes in radiative fluxes is further complicated by the irregular variations that are generated internally. Weather forecasters have recognized that the chaotic behavior of the atmosphere results in an exponential decrease in predictability over time—thus storm behavior cannot be accurately predicted beyond a week or two. Such chaotic behavior within the climate system also introduces internal low-frequency variabilities over decades and longer. These internal variations further modify and complicate the response of the system to external forcing and therefore increase the difficulty of both predicting and detecting such response.

Climate change caused by an external forcing (e.g., an enhanced greenhouse effect because of human activities) can also influence the internal variability of the system. Furthermore, small changes in climate variability can cause relatively large changes in the frequency and severity of extreme climate events. So some of the greatest hazards of global warming may be the result of shifts not in climate per se but in its variability—impacts that, so far, are very difficult to predict.

Climate Change and Natural Hazards. Climate can be represented as a probability function or a histogram of the frequency with which weather events occur. A variety of probability distributions is possible, depending on the weather event selected and the location. These distributions are characterized by their means and variances. Some events are "hybrids" and result from the convergence of more or less unrelated conditions. An example was Hurricane Hazel, a tropical storm that gained new energy when it was picked up by a strong midlatitude jet stream.

There are several ways that climate can change. First, the mean of a distribution could shift in one direction or another. An example of this is the global warming of around 0.5°C over the past century. Second, the variance of the distribution could change. Examples of this are decreased variability in the daily temperature range over the United States, the former Soviet Union, and China or the increased variability of rainfall over the Sahel. Third, a distribution could become more or less skewed—for example, a decrease in the number of mild storms but an increase in intense storms. Fourth, the incidence of hybrid events could change. For example, if the midlatitude jet stream migrates northward in a warmer climate, hybrid events like Hurricane Hazel may become less likely.

Rates of climate change can be linear or nonlinear. A sudden change can occur as a result of the superposition of more than one forcing mechanism or because the climate system reaches a bifurcation point and switches into a different stable mode. An example of the latter would be a switch in the ocean circulation patterns from the current "conveyer belt" to one with greater symmetry between the Atlantic and Pacific circulations. There is mounting evidence that climate tends to exist in several states or modes and switches rapidly between them.

Although a change in mean, such as average precipitation, represents a change from what society expects and is thus highly visible, the greatest impact of climate change on human beings probably will be an increase or decrease in the frequency of different types of extreme events. Current thinking about the impact of climate change on several types of

A tornado looms over the horizon of a North Dakota wheat field. Photograph by Eric Meola, copyright 1999, Image Bank.

extreme events is summarized below (see White and Etkin, 1997, for a more in-depth review).

It has been suggested that tropical cyclones and hurricanes might be more frequent or intense in a warmer climate. Arguments for higher frequencies rest on predictions of increasing sea surface temperatures or ocean thermal energy. But global climate modeling based on an assumption of a world with doubled carbon dioxide levels has produced varied conclusions. Thus, there is no persuasive evidence for a significant change in the frequency of tropical cyclones as the planet warms.

Evidence on how storminess will change in a warmer climate is conflicting. It has commonly been argued that since polar latitudes are expected to warm more than mid- or tropical latitudes, the decreased temperature gradient will result in weaker midlatitude storms. This conclusion is complicated by the prediction of an increased temperature gradient in the upper troposphere that would result from a large warming in the tropical upper troposphere. The effect of moisture further complicates the issue. Latent heat release should increase in a warmer climate, thus strengthening storms. But the transport of latent heat in large-scale eddies from the tropics to the poles may reduce the requirement for balancing the global energy budget, which is the ultimate driving force behind midlatitude storms. Changes in the circulation pattern may well alter storm tracks, an effect that could swamp other regional considerations.

Convective storms are severe thunderstorms that produce hail, lightning, tornadoes, heavy rain, and strong winds. More frequent and more intense convective storms would be expected to result from climate warming because a warming surface and a cooling stratosphere in midlatitudes should destabilize the troposphere. Research seems to bear this out. Recent global circulation modeling (GCM) studies suggest that more frequent and intense convective rainfall would occur in a warmer climate; thus, the frequency of both floods and droughts may increase. But the issue of dynamics also needs to be addressed, and it is not clear how changing wind fields will affect trigger mechanisms. A correlation has been demonstrated between annual temperature and disastrous convective storms and winter storms in the United States from 1950 through 1989. The extreme weather occurred mainly in the southern and eastern United States.

In a warmer climate, heat waves would become more frequent while cold waves would become less so. Evidence suggests that even a warming of a couple of degrees Celsius could result in a major increase in heat waves. The research record on this topic provides relatively consistent evidence. In addition, some researchers have concluded that extreme events are even more sensitive to variability than to average temperatures. This means that many GCM studies may have underestimated the likelihood of extreme temperature occurrences.

Warmer atmospheres can hold more moisture, so precipitation would increase a doubled-carbon dioxide world. Precipitation also is expected to become more convective in nature. Climate change models point with high confidence to an increase in floods as a result of a trend toward more convective precipitation and greater atmospheric absolute humidity. Rising sea levels will likely worsen the impact of storm surges.

If precipitation becomes more convective with an increase in heavier events, the number of dry days should increase and drought will become more severe. This could be exacerbated by increases in potential evapotranspiration because of higher temperatures. Although the evidence is still not clear, there is a reasonable basis for concern that the frequency of prolonged meteorological droughts will increase in a warming climate.

Various other hazards are tied to the primary ones discussed above. For example, wildfires are a function of temperature, the precipitation regime, and lightning. Thus, a longer and more severe forest fire season may accompany climate warming. Storm surges and waves result from ocean- or lake-based storms. It is not clear how these hazards will evolve if the climate warms.

Decisionmaking. The complexity of climate forecasting and the uncertainties inherent in prediction present a dilemma for decisionmakers. On the one hand, such uncertainties suggest that decisionmaking in response to the threat of climate change is premature. However, because of the long time horizons for both the human influence on climate and the adaptability of society to expected change, it may also be risky to delay.

One of the key dilemmas in dealing with complex systems like those described in this chapter is the challenge of making the link between long- and short-run changes. Decisionmaking with regard to climate change illustrates well two types of errors commonly used in statistics. A "Type 1 error" is a situation in which policymakers profess (and perhaps even act on) a belief that there has been a structural change in a system (say a change in climate) but in fact there has not really been one. In a Type 2 error the opposite happens: a structural change in a risk-producing system does occur, but the policymaker believes that there has not been one and therefore takes no action. In the case of climate change the costs of making a Type 2 error are much higher than the costs of making a Type 1 error, so it may be better to take predictions of scientists seriously even if there is considerable uncertainty. To make matters more difficult, the costs of Type 1 and Type 2 errors may be different for different stakeholders—an immensely complicating factor for elected policymakers.

Recent assessments of climate change, particularly those coordinated by the Intergovernmental Panel on Climate Change (1995), have attempted to define what can and cannot be concluded at this time, in terms useful for decisionmakers, as follows:

> There is high confidence in the conclusion that human activities have significantly altered those components of the atmosphere that affect its insulating properties and have therefore enhanced the natural greenhouse effect. These changes will further increase as human emissions of greenhouse gases continue. There is a real and significant chance that such changes will cause global climate changes that are dangerous to ecosystems and human society.

> The balance of evidence from analyses of past climates suggests that there already exists a discernible human influence on the climate system.
>
> Prediction with respect to the magnitude, rate, and regional characteristics of future climate change remain largely uncertain, although broad conclusions with respect to continental and latitudinal responses can be made with some confidence.
>
> In general, predictions of changes in the probabilities and intensities of extreme events as a result of an enhanced greenhouse effect remain unreliable, although there are indications that more rainfall will occur as extreme events and that droughts also may become more frequent and severe.

Parties to the United Nations Framework Convention on Climate Change (FCCC) committed to undertake national programs to mitigate the negative effects and maximize the benefits of climate change, and the FCCC specified that all parties should cooperate in getting ready to adapt to the effects of climate change. The FCCC also acknowledged that major uncertainties precluded full understanding of the problem, that a clear definition of the full extent of the danger and consequences is not yet possible, and that the consequences of climate change could include catastrophic disruptions of climate in some regions.

Climate change is expected to have, overall, an adverse effect on both human systems and the earth's ecosystems. This is because, over a long period of time, humans have adapted to the climate that exists now, not the one expected in the future. Sidebar 4.1 shows the estimated adaptability of various human activities to climate change. In part, the adverse impacts will be reduced by some unknown amount as humans adapt to them. Although society has had a great deal of experience adapting to climate through, for example, the use of building codes, design practices, and changing crops, adapting to fairly rapid climate change is a different story. Research on social adaptation to climate change is needed.

In future decades, as a better understanding of the significance and hazards of climate change is reached, aggressive mitigative action likely will be necessary. Scientists have already indicated that global emissions well below current levels will eventually be required to stabilize concentrations at twice today's level. Technological advancement and social adjustments may also increase the range of measures that may be considered cost effective and hence precautionary. The international community is committed to regular scientific assessments and policy reviews.

Earthquakes in California

Southern California has had more than 200,000 earthquakes in the past decade. A few were damaging, but most were not even felt. Not one of these earthquakes was on the San Andreas fault—the source of all faulting in Southern California. In fact, the last earthquake on that fault's southern stretch was in 1857.

Earthquakes in California are caused by the movement of huge blocks of the earth's crust. Southern California straddles the boundary between the Pacific and North American plates. The Pacific plate is moving northwest, scraping horizontally past North America at a rate of about 1.75 inches a year. About two-thirds of this movement occurs on the San Andreas fault system, which includes the San Andreas, San Jacinto, Elsinore, and Imperial faults. These faults have been studied in greater detail than others because they present a significant hazard to the people who live in southern California and because their high slip rate and behavior make them clearly defined at the surface and easier to study. The faults are so long that they will probably not have one great earth-

SIDEBAR 4.1

Sensitivity and Adaptability of Human Activities and Nature to Climate Change

Activity/Resource	Relationship to Climate Change
Farming	Sensitive but can be adapted at a cost.
Managed forests and grasslands	Sensitive but can be adapted at a cost.
Water resources	Sensitive but can be adapted at a cost.
Tourism and recreation	Sensitive but can be adapted at a cost.
Settlements and coastal structures	Sensitive but can be adapted at a cost.
Human migration	Sensitive but can be adapted at a cost.
Domestic tranquility	Sensitive but can be adapted at a cost.
The natural landscape	Sensitive; adaptation is questionable.
Marine ecosystems	Sensitive; adaptation is questionable.
Industry and energy	Low sensitivity.
Health	Low sensitivity.

SOURCE: Committee on Science, Engineering, and Public Policy (1992).

quake extending down the full length of each fault. Instead, they will probably have smaller earthquakes on shorter segments. These four faults are among the fastest moving, and therefore most dangerous, in southern California.

This is not the whole picture in southern California. The Pacific plate is moving to the northwest, but much of the movement is not parallel to the San Andreas fault. The Pacific and North American plates push into each other, compressing the earth's crust into the mountains of Southern California and producing additional faults and earthquakes.

The rates at which surface faults accumulate slip have been determined by measuring the separation of geological features offset by that fault. Geodesy can also be used to look at how fast faults are accumulating slip. Through the satellites of the global positioning system, the positions of sites can be measured with an accuracy of less than 1 centimeter. This movement must eventually be released by slip on some fault.

The magnitude of an earthquake is a measurement of the energy it produces, not what people or the constructed environment experience during the event. What is experienced is very complex—hard or gentle, long or short, jerky or rolling—and not describable with one number. Aspects of the motion are described by the peak velocity—how fast the ground is moving; the peak acceleration—how quickly the speed of the ground is changing; the frequency—energy is released in waves and the waves vibrate at different frequencies just as sound waves do; and the duration—how long the shaking lasts. Three factors primarily dictate what people feel (and objects endure) in an earthquake: magnitude, distance from the fault, and local soil conditions.

At any instant in an earthquake, at a given distance from the fault, more intense shaking is experienced from a big earthquake than from a small one. Big events also release their energy over a larger area and for a longer period of time. For a magnitude 5.0 event, the actual process of rupturing the fault is over in a few seconds, although shaking might continue for longer because some waves bounce and echo within the earth. Each cycle of shaking stresses buildings and adds to the damage. Because the stress is cumulative, the most severe damage happens at the very end of an earthquake.

Earthquake waves die off as they travel through the earth, so shaking becomes less intense farther from the fault. People near an earthquake will experience all the frequencies produced by the earthquake and feel "jolted." Farther away the higher frequencies will have died away and people will feel a rolling motion. The amount of damage done to a building does not depend solely on how hard it is shaken. Different structures respond differently to various frequencies. In general, smaller buildings like houses respond more to the higher frequencies, so proximity to the fault is a very important factor. Larger structures such as high-rise buildings and bridges are more responsive to lower frequencies and will be more noticeably affected by the largest earthquakes.

Certain soils greatly amplify the shaking in an earthquake. Seismic waves travel at different speeds in different types of rock. Passing from rock to soil, the waves slow down but get bigger. A soft loose soil will shake up to 10 times harder than hard rock at the same distance from the same earthquake. Areas covered by recent sediment, such as the Los Angeles basin and the San Fernando and San Gabriel valleys, may experience greater shaking intensities. Several other factors can affect shaking. Earthquake waves do not travel evenly in all directions from the rupture surface; the orientation of the fault and the direction of slip can change the characteristics of the waves in different directions.

Physical scientists have come together as part of the Southern California Earthquake Center to share knowledge and estimate probable earthquakes in southern California. They combined geodetic, geological, and seismic information to estimate frequencies of damaging earthquakes in different zones. They defined seismic hazard by the characteristic earthquake rate, the rate of all distributed events, and the limiting magnitudes of earthquakes determined by fault lengths. Earthquake rates were determined by observed seismicity and seismic moment rates.

Their analysis yielded a prediction that southern California would experience a magnitude 7.0 or greater earthquake about seven times each century (Working Group on California Earthquake Probabilities, 1995). About half of these quakes will be on the San Andreas system and half on other faults. Their models predicted an 80 to 90 percent probability of a 7.0 or greater earthquake in southern California before the year 2024. Stress on the faults in southern California is accumulating and has been since the southern section of the San Andreas broke in 1857.

The probability of a damaging earthquake is also up in northern California. In 1987 the Working Group on California Earthquake Probabilities (organized by the U.S. Geological Survey at the recommendation of the National Earthquake Prediction Evaluation Council) computed long-term probabilities of earthquakes along the San Andreas fault system on the basis of consensus interpretations of information then available. Its 1990 report raised to 67 percent the probability of one or more large earthquakes in the San Francisco Bay area in the coming 30 years, up from the 50 percent probability that had been set before the 1989 Loma Prieta earthquake (Working Group on California Earthquake Probabilities, 1990).

The Human System

The characteristics of people and of the groups in which they live are the second major system that interacts with others to determine disaster losses. If disasters only struck homogeneous households residing in similar physical structures distributed across a uniform geographic surface in a more or less random fashion, losses might simply be a function of the physical agent's magnitude and where it strikes. But that, of course, is not the case. Not only are people different, but they are changing continuously, both as individuals and as groups. This constant change within the human system (e.g., job shifts to service and sales from manufacturing, producing lower salaries; reliance on off-shore production for U.S. firms making them more vulnerable to unmitigated risks in other countries; increasing reliance on high-tech mechanisms for communications and business transactions) interacts with the physical system to make hazard, exposure, and vulnerability all quite dynamic. Several other hazard-significant shifts in the human system are described below.

Demographics

The type and extent of disaster losses that the United States experiences are, in part, a function of its population, including its size, growth, composition, and distribution. As areas become more densely populated, they also can become more exposed to hazards. The permanent population of the hurricane-prone coastal counties of the United States, for example, continues to grow at a rapid rate. Recent demographic trends point toward increasing exposure to many natural hazards.

The U.S. Bureau of the Census reports the current population of the United States at about 249.2 million and the annual growth rate at about 0.8 percent, of which an increasing portion is the result of immigration. A midrange Census Bureau estimate puts the U.S. population at 288 million by the year 2006.

Of greater importance to disaster exposure is how the population is dispersed on the land. The movement from rural to urban areas continued at a rate of about 1.3 percent between 1970 and 1990 (World Resources Institute, 1992). Almost half of the world's 6.2 billion people will live in urban settings in the year 2000; by 2150 the figure is expected to reach 80 percent. The result of urbanization is an increase in localized population density, heightening the probability that even small-scale disasters will affect larger numbers of people. Moreover, most of this urban group will reside in megacities—giant urban complexes that are contiguous and interdependent and that sometimes extend beyond political boundaries. Because megacities are a fairly recent phenomenon, human experience with disasters in those sites is thin. But some recent events in urban areas (Kobe earthquake, Hurricane Andrew) have shown that there is enormous potential for catastrophic disasters in megacities and that many of them are becoming increasingly vulnerable (see Parker, 1995).

The location of these urban concentrations is crucial as well. About one-third of the world's urban population lives within 60 kilometers of a coastline, and the proportion is increasing (World Resources Institute, 1992). In the United States about 69 million people, or 27.9 percent, live within 100 miles of a coast. Nearly all of the nation's megacities lie within coastal plains. Coastal areas are exposed to an assortment of natural hazards related to wind and flooding. Not only is more U.S. coastline being developed, but the development is increasingly dense and capital intensive. Quaint summer cottages in remote beach areas and barrier islands reminiscent of a bygone era are quickly being replaced with high-rise condominiums and expensive office buildings. When this type of development is damaged, the costs are very high. This trend also has tremendous implications for population dislocation.

Another salient population trend is mobility. The U.S. population is on the move, and this has implications for disaster exposure and vulnerability. The extent to which a community is composed of relatively new residents, whether from other countries or elsewhere in the United States, affects the extent of community and household "memory" of past disasters. Migration also affects the extent to which households are linked to family, neighbors, friends, and local institutions—social networks that are important in shaping disaster response and outcomes (see Chapter 5).

Another important population change is the structure of U.S. households. The trend is toward smaller households; their composition is changing as well. Only 26.3 percent of the total housing units in the 1990 census were occupied by married couples with minor children. Single-person and single-parent households are increasing. While these smaller households may be more mobile in times of disaster, they are likely to have limited economic and human resources to respond effectively.

A related element of urban life is increased interdependency. Most households depend for their daily needs on an extensive network of private and public providers of services and goods. Therefore, what impacts one segment of the population is likely to affect others. When the infrastructure of a community suffers large losses in a disaster, even households that escape major property damage will have a higher probability of encountering hardships in meeting their routine needs and thus will increasingly need to draw on government resources.

These demographic trends will alter exposure and vulnerability to hazards by increasing the number of households at risk for major disasters; increasing the impact of geographically small-scale events; increasing population dislocation after disasters; increasing the necessity for cooperation among contiguous incorporated areas after future disasters; resulting in fewer "local disaster experienced" people in high-risk communities; and resulting in less reliance on kinship and other personal networks for aid in times of disaster and greater need for an increased governmental response.

A full understanding of these trends requires further consideration of human populations and their organization in terms of hazards. For example, although increased population and increased density suggest increased exposure, for some hazards loss of life has declined instead. As noted in Chapter 3, the trend in loss of life from hurricanes in the United States since the turn of the century is clearly downward, even though the population has grown significantly. Average economic losses from hurricanes, however, have increased. The implication is that the relationship between population and loss is not straightforward but rather is complicated by other factors. More than demographics must be considered.

Culture

Shared customs, beliefs, values, knowledge, and skills that guide a people's behavior are their culture. Cultural perceptions and forms of social organization influence exposure to natural hazards. For example, selective attention to risk and preferences for taking or avoiding certain types of risk correspond to cultural biases; and cultural values and beliefs can retard change by defending the status quo.

Vulnerability is in part a function of what technology is utilized and how it is distributed. Those decisions are made by often competing interests, and a host of political, social, and economic factors come into play. Other factors are religion and the ideologies a group uses to justify its internal relationships and its relation to the natural environment. Each of these factors is itself multifaceted and interdependent.

Politics

Public policies and political arrangements also influence the exposure and vulnerability of people to hazards, the impacts experienced, and the effectiveness of the recovery. For example, turf battles can hinder effective interorganizational and international planning. By the same token, the emergence of fresh postdisaster alliances between regions, public/private-sector groups, and others can overcome the weaknesses of traditional reconstruction practices.

Social Characteristics

The different economic, cultural, and other characteristics of individuals and groups in society affect the distribution of both exposure and vulnerability to hazards and disaster impacts. It may well be that some of the nation's past and current efforts to mitigate exposure to hazards and vulnerability have inadvertently redistributed it away from some people and toward others, rather than actually reducing it.

A major approach to hazards research over the past two decades has looked at the way in which a variety of socioeconomic characteristics of people affect (or "structure") their vulnerability to hazards and disasters over time. Known as the social vulnerability paradigm, in this perspective certain people experience socially created vulnerability—an elevated probability of loss, injury, death, and/or reduced ability to recover from hazards or disasters that is due to a range of social, political, and economic processes (see Hewitt, 1983, 1997; Blaikie et al., 1994; Peacock et al., 1997). This approach has become a mainstream paradigm in much of Latin America, Australia, southern Africa, and South Asia and has been adopted by many international organizations such as the International Federation of Red Cross and Red Crescent Societies, the World Health Organization, the United Nations University, the Church World Service, and Oxfam.

In the United States the key characteristics that seem to influence disaster vulnerability most are socioeconomic status, gender, and race or ethnicity. Differences in these characteristics result in a complicated system of stratification of wealth, power, and status. This stratification, in turn, results in an uneven distribution of exposure and vulnerability to hazards, disaster losses, and other impacts and access to aid, recovery, and reconstruction. For example, the poor are more likely to occupy old and more hazardous housing, ethnic minorities are less likely to receive disaster warnings and are more likely to have language barriers to the information they receive, and developing nations are less able to afford disaster detection technology and resilient construction practices.

As used here, socioeconomic status means a category of people who have about the same amount of income, power, and prestige. In 1994 the bottom 20 percent of the nation's people had a 3.6 percent share of total income, the middle 60 percent had 47.3 percent, and the top 5 percent held 21.2 percent of the total income; 38.1 million people lived at or below the poverty line. The gap between wealthy and poor in the United States continues to grow, while the middle class continues to shrink.

Note that in many (but not all) instances, gender, race, and ethnicity are not the key factors in increased exposure or vulnerability but rather are indicators of lower socioeconomic status and a relative lack of power, which are key factors. Census data show that people of color are disproportionately poor and that the proportion of poor among racial and ethnic minorities in the United States is growing. Women around the world are in the lower ranks of economic stratification. In 1991 female heads of households and their children accounted for 52 percent of the poor in the United States. Women who are members of ethnic/racial minorities are even more likely to experience poverty.

The poor are at greater risk from natural and technological disasters in the United States and worldwide mainly because they live in lower-quality housing that is more likely to be damaged and often is closer to technologically hazardous sites. The relationship between housing vulnerability and race/ethnicity exists in the United States because of segregated residential patterns and economic barriers to sound construction. Desirable beachfront property is an exception. Affluent whites are more likely to live there and to insure themselves against damage. Exposure for the poor is even greater in developing nations, where housing density in hazardous areas is greater and there is less choice about where to live.

Globally and in the United States, women are more vulnerable to disasters than men. One reason is that women are disproportionately poor and more vulnerable in general. But women's vulnerability is also influenced by their roles: they are more likely to remain with family members in emergencies to nurture, assist, and protect them.

The uneven distribution of exposure and vulnerability among people is borne out when disasters strike. Low-income groups—largely minorities and women—have the highest disaster mortality and morbidity rates. The nation's elderly poor are the most likely to die in heat waves. Of the 148 deaths in the 1980 Midwest heat disaster, most were elderly and poor; many of these people worried about high utility bills and so did not turn on their fans. Mobile homes, often occupied by lower socioeconomic groups, are the most dangerous housing type in tornadoes; in 1994 almost 40 percent of all tornado fatalities occurred in mobile homes (U.S. Department of Commerce, 1995). Unreinforced masonry buildings typically house the poor, and they are the most likely to collapse in earthquakes. Anyone who is poor in the United States is more likely to become homeless after a disaster. Although it is known that men are more likely to be killed by lightning and that more women die in tornadoes, mortality rates by gender in this nation are not well documented. Research in other nations shows that women and children often account for a disproportionate number of deaths from disasters.

Emotional stress, trauma, and other psychological impacts of disasters also are unevenly distributed in the United States. Higher-income victims suffer far fewer psychological impacts than do low-income victims, largely because disasters exacerbate poverty. The relationship between socioeconomic status and psychological impacts also holds in the developing world. Research findings on gender and psychological impacts are mixed but in general show that women and girls suffer more emotional problems after disasters, while men experience greater decreases in mental and physical well-being and have increased rates of depression and alcohol abuse. Women often suffer from physical and emotional exhaustion in the posttrauma period.

Research shows that people of lower socioeconomic status have the most trouble reconstructing their lives and reestablishing permanent housing after disasters in the United States. They have less insurance, more financial stress, more trouble negotiating bureaucracies, less access to resources, and more difficulty obtaining loans. Housing services in the United States are geared toward homeowners, and many areas simply lack affordable housing even before a disaster strikes.

The socioeconomic, racial, and ethnic composition of the United States is not static. The number of poor is on the rise, and the nation is becoming more racially and ethnically diverse. Income disparity—the decline of the middle class and the evolution of the nation into a two-class system of rich and poor—continues to grow in almost every state. The process is strongest and fastest along the nation's coasts and in major urban areas—the very places that will face more natural disasters in the future.

In 1994, 38.1 million people lived at or below the poverty line in the United States. By 1995 the figure was 39.2 million. But in 1996 upper income families increased their incomes by 2.2 percent. The proportion of poor among racial and ethnic minorities in this nation—already high—is growing. And the nation is becoming more racially and ethnically diverse. The Census Bureau reports that in 1994, 74 percent of the population was white/non-Hispanic, 12 percent was black, 10 percent was Hispanic, and 4 percent was Asian/other. Census Bureau projections show a dramatic shift: in 2050, 53 percent of the U.S. population will be white/non-Hispanic, 22 percent will be Hispanic, 14 percent will be Black, and 11 percent will be Asian and other. Shifts in the populations of the three scenario cities examined in the first assessment (White and Haas, 1975) and in Chapter 2 here illustrate the dynamic nature of these characteristics (see Sidebar 4.2).

The different socially based levels of vulnerability described above have not always been taken into consideration in past attempts to cope with and minimize disaster and hazard losses in the United States. Knowing about these different levels of susceptibility to loss and impacts and their interactions with aspects of the physical systems (and the constructed environment discussed in the next section) will help target future research and management efforts. All of these characteristics are constantly in a state of flux. Furthermore, some kinds of social vulnerability are growing and thus will increase exposure and vulnerability to hazards in the future. A future course toward sustainable hazards mitigation, with its emphasis on equity (as described in Chapter 1), should have as one goal helping to alleviate this inequality and giving more people better opportunities to be and stay safe.

Economics

Besides the tremendous influence that socioeconomic status plays in the distribution of hazards-related vulnerability, economic interests and systems also have dramatic implications for exposure to natural hazards. Consumption practices in wealthy nations relegate environmental management to the back seat as vulnerable regions become densely developed and populated. In developing countries, rapid and unplanned urbanization, subsistence living, and economic development contribute to soil erosion, desertification, and increased vulnerability to disasters.

Additionally, the growing economic interdependence of the world ensures that local disasters can have increasingly global impacts in the future.

Special Groups

Approximately 25 percent of the nation's citizens live in rural areas. Rural areas tend to have lower average incomes and higher poverty rates; many also have high proportions of children and elderly. Rural areas typically have a lower resource and tax base and fewer organizations, businesses, specialized services, and resources. Rural areas have a stronger sense of community than urban areas. Power is often controlled by a few, and outsiders are often treated with suspicion. Little is known about

SIDEBAR 4.2

Demographic Shifts in the Three Scenario Cities, 1970 to 2015

Dade County, Florida

In each year of the 1990s, Dade County, Florida, added 39,000 people to its population. This population growth was due largely to international immigration, mostly Hispanic. In 1960 South Florida's population was 1.3 million; by 1990 it was 3.3 million. Population predictions show South Florida with 3.7 million residents by the turn of the century and 4.3 million residents by 2015. In 1970 Dade County had a population of 1,267,800; in 1980 it increased to 1,625,800; by 1995 the population was 2,057,000. Projections show the Dade County population to be 2,291,000 in 2000 and 2,775,000 by 2010. Moreover, Miami and Dade County are increasingly racially and ethnically diverse. In 1990 Dade County was 72.9 percent White, 20.6 percent Black, 6.5 percent other, and 49.2 percent Hispanic (of any race). By 1997, the percentage of Hispanics in Dade County was over 50 percent. In the city of Miami in 1990 the population was 65.6 percent White, 27.4 percent Black, 7 percent other, and 62.5 percent Hispanic (of any race).

San Francisco Bay Area

In 1970 the San Francisco Bay area had a population of 4,638,800; by 1995 the population had increased to 6,484,300. Between 1970 and 1994 the area grew by 38.5 percent, an increase due largely to foreign immigration, mostly Hispanic. Between 1995 and 2015 the bay area will add about 1.2 million new residents. In 1990 the nine-county Bay Area was 68.9 percent White; 8.9 percent Black; 0.6 percent American

the particular vulnerability or response of rural areas to disasters. It may be substantially different from that in urban settings.

When the Americans with Disabilities Act was enacted in 1990, some 17 percent of the nation's population was considered disabled, and it was estimated that one-third of these people were at least 65 years old. Researchers agree that disabilities affect how people cope with exposure to and the impacts of disasters, but there has not been much research on the topic. Different strategies are needed for those with impaired vision, hearing, mobility, mental processes, and combinations of those. The Federal Emergency Management Agency, the American Red Cross, the city of San Francisco, and others have taken some steps to understand and help disabled people prepare for and cope with emergencies.

Indian, Eskimo, or Aleut; 15.3 percent Asian or Pacific Islander; and 15 percent Hispanic. In 1990 there were 549,190 people in the bay area who lived in households that receive public assistance. In the city of San Francisco there were 678,974 residents in 1980; by 1990, the population was up to 723,959. In 1990 there were 378,783 Whites, 79,039 Blacks, 207,155 Asians, and 100,717 Hispanics.

Boulder County, Colorado

In 1970 Boulder County, Colorado, had 131,889 residents, and in 1990 its population had increased to 238,196. In 1970 the Black population in the county was 0.5 percent. In 1980 the county was 98.4 percent White, 0.93 percent Black, .49 percent Native American, 1.29 percent Asian, and 5.35 percent of Spanish origin. In 1990 Boulder County was 88.24 percent White, 0.82 percent Black, 0.55 percent Native American, 2.31 percent Asian, and 6.38 percent of Spanish origin. Thus, the nonwhite population has grown slightly in the county. In 1980 10.1 percent of the county's residents lived below the poverty line; in 1990 11 percent did. In 1970 the median family income was $11,000; in 1980 it was $23,705; in 1990 it was $35,322. In 1990, 3,096 families (5.6 percent) in Boulder County lived below the poverty level. In 1980 in the city of Boulder there were 76,685 residents; by 1997 there were 95,662 residents. In 1980 Boulder had 11,353 people living below the poverty level; in 1990 there were 14,393 people below the poverty line. In 1990 there were 1,196 families (7.5 percent) in Boulder living below the poverty line. In 1980 Whites were 94.5 percent of the population of Boulder, Blacks were 1.3 percent, and those with Spanish origin made up 3.7 percent. In 1990 Whites were 92.5 percent, Blacks were 1.2 percent, and those with Spanish origin were 4.8 percent.

Our nation's aging population and increased numbers of people otherwise without the ability to respond quickly or on their own to an emergency suggest that the nation will have a much larger "dependent-in-disasters" population in the future. There is a need for research into these groups' needs, how to provide for them, and how to include them in sustainable hazards mitigation.

The Constructed Environment

The third major system in the structure of hazard is that of the human-made environment and technology—public utilities, transportation systems, communications, critical facilities, engineered structures, and housing. The ability of this built environment to withstand the impacts of extreme natural forces plays a direct role in determining the number of lives lost, the number and severity of injuries, and the financial impact of disasters. Building codes and wise construction practices (which are discussed in Chapter 6) can strengthen the nation's constructed environment. The building codes that are in place, the extent to which they take hazard mitigation into account, how well they are enforced, and the construction practices that follow from the codes all define hazard and future losses suffered by structures. The types and extent of losses suffered in Hurricane Andrew were partly a result of the codes and practices in place in Southern Florida in 1992.

Practices and Losses in Hurricane Andrew

Hurricane Andrew was one of the most thoroughly investigated storms in the history of the United States because of its severity and the incidence of major shortcomings in construction techniques and code enforcement. Reports after the hurricane suggest that wind speeds did not greatly exceed 120 mph. The South Florida Building Code uses the 120 mph benchmark in establishing design requirements.

Field investigations of damage revealed that loss of roof cladding was the most pervasive type of damage to buildings in southern Dade County. All kinds of roofs suffered damage because of failure in the method of attachment and materials, inadequate design, poor workmanship, and missile or debris impact. The loss of roof material and roof sheathing and the failures of windows and doors exposed the interiors of buildings to more damage from wind and rain, making thousands of homes and businesses uninhabitable.

Loss of shingles, tiles, and siding not only compromised the integrity of the buildings, but these objects also became wind-borne debris that damaged other buildings and automobiles. Failure of gable ends was the most frequent residential structural failure observed. The South Florida Building Code requires the use of construction straps, and many builders did attempt to comply with the codes. However, in many instances they were improperly installed or the wrong materials were used.

In economic terms, one of the most damaging failures was not structural but rather was set in motion by loss of gypsum wallboard ceiling. This damage affected many homes that were not even structurally damaged. Rain was driven in through gable-end vents, through the joints between roofing panels after tile or shingles were blown off, and directly through the attics under failed roofing systems. Rain saturated the insulation as well as the ceiling, leading to widespread collapse of ceilings. This pattern illustrates the importance of roof coverings, which generally undergo very little testing for hurricane resistance. The failure of asphalt shingle roofs was usually related to fastener attachments tearing through the shingles. Not surprisingly, stapled attachments were more likely to be torn off by the wind, even though the staples seemed to perform well under laboratory conditions.

Inadequate design for load transfer of wind forces was found to be the major cause of structural failures. Proper connections between critical components allow for the safe transfer of wind loads required to maintain structural stability. Where high-quality workmanship was present, the building performance was significantly better (see Federal Emergency Management Agency, 1992). It was clear that not all tradespeople were qualified to build the structural systems, structural components, and connections necessary to resist design wind loads.

Inadequate county review of construction permit documents, organizational deficiencies such as a shortage of inspectors and supervisors, and inadequate training of those personnel may have contributed to the poor quality of the construction. The problem of detecting substandard workmanship was magnified in large developments of similarly designed homes because of the repetitive nature of the inspection process.

Where one- and two-story wood-frame buildings structurally failed, investigators found improper installation of framing connections and an absence of load transfer straps or bracing from nonload-bearing walls to connecting walls and roof components, among other shortcomings. Less structural damage was observed in modular housing. The module-to-module combination of those units appears to have provided an inherently rigid system that performed better than conventional residential framing. For masonry homes the lack of vertical wall reinforcing was a major cause of failures. However, because masonry walls have a much heavier mass and such structures employ a construction system less prone to failure from design and construction errors, masonry construction generally performed better than all-wood-frame construction.

Mobile homes suffered damage far out of proportion to that of other construction. In several mobile home parks in southern Dade County all units were total losses. The debris remnants of such homes often posed a substantial hazard to nearby buildings. Mobile home construction is generally guided by federal standards, which have had relatively low design wind-speed requirements. However, the tiedown of a mobile home to its pad is regulated by local building codes. None of the tiedowns observed in the investigations were certified to be in accordance with code requirements.

Lax zoning, inspection, and building codes contributed to the destruction caused by Hurricane Andrew. Southern Dade County homes built after 1980 came through the storm in worse shape than older homes. In heavily damaged areas of southern Dade County, where sustained wind speeds were believed to be less than 97 mph, only 10 percent of the homes built before 1980 had enough damage to be uninhabitable, compared to 33 percent of those built after 1980 (*The Miami Herald,* 1992).

One builder in southern Dade County built homes from 1989 to 1990 that survived the storm virtually intact in the midst of more expensive subdivisions that were heavily damaged. Raul Munne built the 71-home Munne Estates counter to several trends in style, construction practices, and quality control during the 1980s. First, he built concrete block homes with modest roof lines, a building style more suitable to high-wind regions. This style contrasts with the 1980s trend toward wood-frame houses with steep gables. The use of 5/8-inch plywood, nails driven by hand, continuous strips of mortar to hold roof tiles, and careful supervision during the construction process were all cited as reasons why Munne homes survived so well.

Another example of quality building in southern Dade County came from homes built by Habitat for Humanity, a nonprofit organization that uses volunteer labor to build modest homes for low-income families. Six homes built by Habitat for Humanity were located in the vicinity of Hurricane Andrew's highest winds, and all survived with little or no damage. In fact, all 27 Habitat for Humanity homes in Dade County were intact and habitable after the storm. Habitat for Humanity also had used materials and construction techniques well suited to hurricanes.

Despite Hurricane Andrew's devastation, many professionals still view the South Florida Building Code as one of the best for hurricane resistance, even though it has been weakened by modifications and is not completely up to date on national wind-loading standards. Many of the field investigations showed that damage was caused not so much by inadequacies in the code as by poor workmanship, construction deficiencies, building styles not suited to high winds, poor code enforcement, and inadequate inspections. A review of a 35-year period of decisions by the county's Board of Rules and Appeals found a number of instances in which changes were made under pressure from builders and other groups in the name of construction cost savings. Some examples follow:

The code requires buildings to be constructed to withstand hurricane winds of 120 mph, but the board allowed builders to use asphalt shingles on roofs despite an awareness that shingles can only resist sustained speeds of 63 mph.

To save costs, in 1961 the board began to allow builders to use staples instead of nails to install roofs. By 1983 the board learned that staples were not performing well in relatively modest hurricanes and tropical storms, but the code was not changed until after Hurricane Andrew.

Beginning in 1972 the board began to consider revising the code to include workmanship standards. Changes were not implemented until 1992, mostly in response to a critical grand jury report.

In 1980 the board endorsed the use of thinner roofing felt to save costs.

In 1984 the board allowed waferboard (pressed particles of wood) to be used instead of plywood on roofs. It was reported that officials in Dade County's product control division had some misgivings about the alternative materials but felt powerless to question industry-sponsored tests. Waferboard was banned after Hurricane Andrew because of concerns about its performance when saturated with water.

Inadequate staffing was clearly a problem in Dade County even though the county has more resources and a history of more intense development than many coastal counties. Nevertheless, at the time of Hurricane Andrew, there were only 60 building inspectors for all of Dade County. They were required to inspect an average of 20,000 new buildings each year, or about 35 inspections per day for each inspector—a nearly impossible feat. This state of affairs cannot be blamed entirely on the building department. The politics, the budgetary process and pressures, and the public's relatively low awareness of the importance of building codes and construction practices also played a role. Many municipalities look at permit and inspection fees as a revenue producer for other departments, overlooking the fact that building departments have an important mission in protecting citizens' health, safety, and general welfare.

Implications

The findings from Hurricane Andrew illustrate well the interaction between the physical system (as manifested by the hurricane) and the constructed environment of parts of south Florida. The magnitude of the losses was in many cases a function of misunderstanding of, or failure to address, the impacts of such interactions.

The affordability of mobile homes makes them the fastest-growing segment of the nation's housing stock. In fact, these homes now account for half of new single-family homes purchased each year in the South. But their performance in Hurricane Andrew shows that more care is needed in specifying their wind resistance. In 1994 the U.S. Department of Housing and Urban Development announced tougher standards for mobile home construction in hurricane-prone regions of the country. Mobile homes sold in Hawaii, coastal Alaska, and 25 coastal counties in Louisiana, Florida, and North Carolina now have to withstand winds of 110 mph. Such homes sold in 11 other hurricane-prone states will have to resist 100 mph winds. The rest of the United States will be guided by an 80 mph standard.

Many field investigations showed that the damage from Hurricane Andrew was caused by poor workmanship, construction deficiencies, building styles not suited to high winds, poor code enforcement, and inadequate inspections. The desire to reduce costs was a driving force in some cases. Affordable housing is a worthy goal, but it must be balanced with other objectives. Some code officials argue that the building codes need more advocates for safety and loss mitigation, including consumer groups, emergency management organizations, and insurers. These groups can be a counterweight to builders and materials manufacturers interested in lowering costs or promoting new products. Building styles appropriate to likely hazards, high-quality construction and workmanship, and proper inspections are all needed if the constructed environment is to avoid being the "weak link" in disaster situations.

THE HAZARDOUSNESS OF THE NATION

Disaster losses are the result of the interaction of and changes in the characteristics of the physical environmental systems that produce extreme events, the characteristics of the people and communities that experience those events, and the nature of the constructed environment that is affected. The United States is probably facing a future more hazardous than its past. There is a greater and growing chance of larger losses associated with natural and related technological hazards and disasters. This judgment rests on the changes that have been observed in the three systems that interact to result in disasters. The physical systems that give rise to extreme events are complex and changing. A warming climate alone is expected to result in more extreme meteorological events. Changes in the composition and distribution of the population are redistributing larger numbers of people to more hazardous areas. The socioeconomic characteristics of the nation's people are changing, resulting in a larger proportion with low incomes and thus more likely to be seriously affected by disasters. And the codes and construction practices being used in hazardous areas are not necessarily reducing the vulnerability of the nation's built environment.

This chapter brought together only a few elements of the physical, human, and constructed systems that shape hazard. Not enough is known about changes in the physical, social, and constructed systems of the United States that are reshaping its future regarding hazards. Nor are the ways in which those systems interact to shape future losses well understood. To remedy this deficiency, a national risk assessment should be undertaken immediately. It should blend a comprehensive treatment of the three sets of factors discussed in this chapter so that hazards can be estimated interactively and comprehensively. The assessment should be crafted so that it supports local actions for sustainable hazards mitigation. It should be end to end; that is, from the largest relevant global physical processes and changes down through their hazards and implications for the people, resources, buildings, structures, and decisionmaking in the nation's local communities, and vice versa. Many other topics should be examined in a thorough national risk assessment: the emergence and growth of global economic systems and their linkages, deforestation, the aging of the nation's population and infrastructure, and the many other factors that contribute to the nation's hazardousness.

CHAPTER FIVE

Influences on the Adoption and Implementation of Mitigation

IN AN IDEAL WORLD, self-reliant households, businesses, and organizations would take adequate steps to avoid, mitigate, or otherwise cope with natural and related technological hazards if left to their own devices. Those exposed to such hazards would accurately assess their risk, search for possible adjustments, and adopt techniques that would provide them with adequate protection at a reasonable cost. They would avoid hazardous locations or build hazard-resistant structures, learn how to turn off utilities to prevent fires, and set aside financial reserves or purchase insurance to guard against financial loss. Residual losses would be borne as a reduction in the standard of living until the victims had fully recovered. It follows that government attempts to regulate choices would be an unwarranted intrusion.

But it is well known and documented that there are many conditions and circumstances that have either positive or negative impacts on people's tendencies to carry out activities that will minimize their chances of loss, injury, or death in hazardous situations. In fact, individuals, organizations, businesses, and governments tend not to adopt or implement on any large scale the mix of sustainable mitigation precautions that would enable them to avoid long-term losses from hazards. Thus, making mitigation a reality will require overcoming many human behaviors along with financial, political, and social obstacles. Knowledge and proven techniques exist to overcome or circumvent some of the obstacles, but others remain mysterious.

Several of these factors were identified under the traditional hazards paradigm as "constraints to adjustment." Over the past two decades a fuller understanding of some of them has been gained, and additional ones have been added. There are also now available ideas, theories, and evidence about what induces mitigation behavior. Five broad categories of influences on human choice and action are discussed in this chapter: the decisionmaking process itself and social, economic, legal, and other factors.

DECISIONMAKING PROCESSES

Research in many fields—political science, economics, sociology, psychology, organizational behavior, and social psychology—has identified varied factors that may explain the extent to which people or groups do or do not take precautions against hazards. Decisions to adopt and implement hazards mitigation can be made at personal, organizational, or governmental levels.[1] One set of choices takes place at the individual level: people who are aware of a hazard and of ways to protect themselves from it may or may not choose to take precautionary steps. In the case of organizations or businesses, the decisionmaking process is somewhat different because the context is different. Governments must make the same types of choices, but again the public decision process differs somewhat from the other two.

The processes that underlie adoption and implementation are complex. The people or groups who stand to gain or lose from an adjustment's implementation differ in many ways—for example, in their knowledge, resources, and decision processes. But the overwhelming scientific evidence is that people typically are unaware of the hazards they face, underestimate those of which they are aware, overestimate their ability to cope when disaster strikes, often blame others for their losses,

[1]As used in this chapter, "adopting" hazard mitigation means an initial commitment of resources to taking precautionary measures, while "implementation" means a continuing allocation of resources to follow through on them.

underutilize preimpact hazard strategies, and rely heavily on emergency relief when the need arises.

In some instances, the failure to adjust to hazards may be entirely rational, especially when severe economic constraints are at work. But it is more generally the case that people and organizations with sufficient resources choose not to protect themselves against even relatively high probability hazards. It must be remembered that individuals, besides acting alone, also underlie group and democratic decisions. So the process that a sole person goes through in deciding whether to adopt or implement hazard mitigation has a bearing on most of the other decisionmaking processes.

Individual Decisionmaking

Understanding of individual decisionmaking is far from complete, but some important pieces of information are in hand. Individuals do not process information about hazards or about precautions to them in perfectly rational ways. Individual decisionmakers often assess imperfectly the situations in which they find themselves, do not know the full range of alternative actions or products available to them, and do a poor job of using the information they do have to evaluate likely states of nature and the consequences of their actions. Individuals are typically unable to maximize utility or appreciate the need for additional information. They lack insight and consistency regarding present and future preferences, planning only for the immediate future and forecasting that future mainly on the basis of the immediate past. Furthermore, individuals are subject to the weakness of relying on flawed cognitive heuristics (such as the gambler's fallacy) to help them reach a decision. They are also likely to be distracted by other issues competing for their attention and to be biased by previously held attitudes and beliefs. A number of models or perspectives from different disciplines help explain individual decisionmaking about hazard mitigation adoption and implementation.

Economic Market Processes

Rational economic decisionmaking models assume perfect markets and perfect information. But in the real world individual decisionmakers are faced with sparse information about hazards and about the probable effectiveness of proposed mitigation. Since decisionmakers have limited information and deal in markets that do not perform like models, the prospect of natural hazards in a specific locale has somewhat unpredictable effects.

Markets related to natural hazards provide imperfect signals because the perceived benefits of effective hazard adjustments tend to be lower than their true social benefits, thus leading to insufficient demand and, consequently, inadequate supply. This is the consequence of decisionmakers' inability to cope rationally with low-probability, high-consequence events. One specific problem is adverse selection in the purchase of insurance. If decisionmakers purchase insurance only because they are at risk, only those most likely to make claims will purchase policies. The inevitable consequence is that insurance companies will fail to collect enough premiums to cover their losses, will raise premiums, or will withdraw from the market (see Kunreuther et al., 1978).

All of these problems contribute to the problematic nature of understanding adoption decisionmaking from the standpoint of an economic model. Few of the economic issues associated with natural hazards appear to be unique to this area; many have been addressed in such areas as occupational safety and environmental pollution. Many issues remain to be resolved, ranging from the quality and price charged by private contractors, and the bargaining power of providers and consumers, through the importance of economic efficiency and social justice in natural hazard adjustment, and the role that government should play in attempting to ameliorate market imperfections.

Utility Theory

Despite its shortcomings in explaining behavior, classical decision theory has had significant impacts on decisionmaking models, such as utility theory. The conditions required for classical utility theory rarely, if ever, exist: full knowledge of alternatives, costs, and objective utility. But its main idea—that a person makes choices in an attempt to "maximize" his or her preferences—is attractive.

Simon's (1957) concept of "bounded rationality" is one such alternative approach (and the one relied on in the traditional paradigm of human adjustment to hazards). It maintains that people are limited to being able to deal with relatively little information and relatively few concepts. Thus, although they cannot be completely rational in terms of classic theory, they can take a rational approach. A second alternative is the idea that people employ subjective estimates of probabilities and of the utility that would be derived from alternative choices as surrogates for the "real" data they do not have.

Neither of these theories has explained the adoption of natural hazards mitigation, however. Bounded rationality leads people to underestimate the risks of natural hazards, which in turn leads to underadjustment, followed by a crisis orientation after disaster does strike. And subjective expected utility theory may work well in static environments but is poorly suited to decisions about environmental extremes, which must be made under conditions of extreme uncertainty. Decisionmakers have a tendency to rely instead on standard operating procedures, incremental changes, and short-term feedback (see Kunreuther et al., 1978).

Heuristics

Another attempt at understanding individual decisionmaking with regard to hazard mitigation is the suggestion that people use heuristics to guide them when faced with complex choices. One, termed "availability," suggests that judgments about the frequency of an event depend on the ease of retrieving specific similar instances from memory. Events that are more easily recalled are judged to be more probable than those that are hard to remember. Thus, for example, local residents may judge that an earthquake is more likely to occur if they have experienced numerous earthquakes in the past or if they suffered severe losses in a recent quake.

Use of an "anchoring and adjustment" heuristic also appears to influence decisionmaking, particularly when people need to make an overall judgment about a body of information that is too big to process at one time. The model suggests that, to integrate the information, decisionmakers begin with an initial estimate based on a limited amount of information and then adjust the estimate as additional information is acquired. Unfortunately, the adjustment to the original estimate is frequently insufficient. This causes little harm when the initial information is the most important but can cause significant bias if the more important information comes later. As an example, suppose that a real estate agent first describes all of the attractions of a home and then discloses its location in a floodplain only late in the presentation. In this situation the anchoring and adjustment heuristic would be expected to lead the prospective purchaser to undervalue the information about the property's hazard exposure.

Prospect Theory

Flaws in human decisionmaking have led to the development of a number of revised theories. One of these, prospect theory, contends that decision problems are solved in phases. In the first phase the decision-maker frames a problem in terms of a relevant set of alternative actions and their consequences. The set of options is edited by translating the potential consequences of a decision into subjective values and the probabilities of those outcomes into decision weights. The subjective values and decision weights are combined in the second phase, while the third phase uses the combined value to produce a decision.

This model has received relatively little attention from hazards researchers except as it pertains to the purchase of insurance. It has been found that individuals tend to make complex tradeoffs between such issues as the probability of an event and its likely outcomes, depending on the context of the problem and the mode in which information is communicated. Moreover, some people have a tendency to treat a very low probability as a zero probability ("it can't happen to me"). Indeed, the failure to purchase earthquake insurance is a specific example, where both costs and benefits are unknown to the decisionmaker. In such conditions, people justify their decisions with arguments that may seem far-fetched or distant from "rational" models of choice (see Hogarth and Kunreuther, 1993).

Attitude Theory

According to this theory, people's behaviors can be predicted from relevant beliefs, values, and attitudes. In turn, their behavioral intentions are determined by their attitudes toward the behavior and their subjective norm for that behavior, including how others would view it (see Fishbein and Stasson, 1990).

One line of research that is applicable to hazards is that dealing with the role of situational contingencies in motivating behavior. Specifically, sensory cues from the physical environment or socially transmitted information (e.g., warnings) elicit a perception of threat, which diverts the recipient's attention from normal activities. Depending on the perceived characteristics of the threat, those exposed to hazards will either resume normal activities, seek additional information, pursue problem-focused actions to protect themselves, or engage in emotion-focused actions to reduce immediate psychological distress. The way an individual chooses to respond to a threat depends on the individual's personal appraisal of both the threat and the available protective actions. Environmental threats tend to be appraised in terms of their certainty, severity, immediacy, and duration, while alternative actions generally are evaluated in terms of their efficacy, cost, time requirements, and implementation barriers (see Perry et al., 1981).

These theories suggest that it is important to assess what people believe about natural hazards and mitigative actions, whether people's beliefs make a difference in adopting and implementing mitigation, and (assuming beliefs do make a difference) how beliefs can be changed to increase the adoption and implementation of effective measures.

Attention to Information

Some psychological research supports the notion that individuals are highly selective in determining the pieces of information to which they will pay attention. They tend to discount or ignore information that conflicts with previously held attitudes. This could be of particular significance to hazards mitigation. For example, a strong personal commitment to living in a freely chosen location would quite likely cause a psychological conflict with new information about a hazard at that location.

Communication Theory

Risk communication is based on the assumption that people make wrong decisions because they are uninformed about the consequences of their actions. Disseminating scientific information would presumably change people's beliefs about a hazard and in turn lead to the adoption of appropriate mitigation strategies. This assumption oversimplifies the adoption process by ignoring variations in the source, channel, and receiver of the message and also the impediments to information processing described above: competing demands for attention, use of cognitive heuristics, and conflicts with existing beliefs.

For example, deciding to take action in response to a hazard warning is a process with several stages: (1) hearing the warning, (2) believing that the warning is credible, (3) confirming that the threat does exist, (4) personalizing the warning and confirming that others are heeding it, (5) determining whether protective action is needed, (6) determining whether protection is feasible, and (7) determining what protective action to take and then taking it. A primary issue in persuasion is the identification of factors that could influence the response at each stage (see Lindell and Perry, 1992).

Researchers have tried to find correlations between hazard adjustment and a broad range of human characteristics, including socioeconomic (age, ethnicity, gender, income, education), geographic (recency and frequency of hazard experience and proximity to the impact area), and psychological characteristics. The results have been mixed. That is, in one instance for one hazard, age could be a determinant of taking protective action, while in another situation community connections are more important. Acceptance of a message depends on its compatibility with the receiver's existing beliefs. A message is likely to be rejected if it conflicts significantly but accepted if the information is novel or poses only a small conflict. Research on warning response has provided evidence that people think of protective responses to disaster in terms of four principal characteristics: efficacy, cost, time requirements, and implementation barriers. These or similar dimensions would be useful in assessing how particular hazard mitigation options would be perceived.

Habit

A persuasive message can achieve behavioral change only if it overcomes the obstacle of existing habits. But the role of habit has received virtually no systematic research attention. Contemporary cognitive theories distinguish between automatic cognitive processes (habits) and controlled processes. Automatic processes are initiated outside conscious awareness by repeatedly encountered situational cues. In the case of natural hazards it is the rarity of actual events that keeps the likelihood of active consideration low. This perspective could be used to determine what new situational cues could be provided to induce new automatic responses to hazards.

Social Expectations

Groups of individuals have social expectations about what should or should not be done in specific situations. These norms become collective habits—the "right thing to do under the circumstances." They are not the outcome of thoughtful decisions intended to be adaptive adjustments to a particular hazard but are the result of people's tendency to conform to the behavior of those around them. This can lead to the adoption of hazard mitigation actions without any awareness of their value in adapting to the physical environment. This behavior is in sharp contrast to the more profound thought processes involved in assessing the expected utility of an adjustment based on the possible negative impacts of the physical environment. This line of reasoning suggests that inducing people to take mitigative action is a problem of overcoming social conformity and encouraging innovation.

Decisionmaking in Organizations

Much is known about the process that formal organizations follow in responding (or failing to respond) to perceived threats from the physical environment. In general, hazard mitigation measures are preceded by awareness of a hazard on the part of the organization, awareness of alternative adjustments, adoption of one or more adjustments, implementation of the adjustments adopted, and, perhaps, subsequent evaluation (see Burton et al., 1978). But knowledge about the factors that influence adoption of hazard mitigation or preparedness measures in organizations is relatively thin.

Decisionmaking in Government

Governments adopt and implement large-scale hazard mitigation directly (such as dams and levees), mandate that lower levels of government and individuals engage in risk reduction activities (such as adopting building codes), and regulate development and land use. A variety of models exist to help understand natural hazard adoption and implementation by governments. Models of organizational decisionmaking are often applicable to public policy decisions. Behavioral models of organizational decisionmaking can help explain municipal adoption of retrofit requirements for unreinforced masonry buildings. Political scientists have long concerned themselves with the processes of governmental policy adoption and implementation, devising behavioral models—political systems theory, group theory, elite theory, institutionalism, and rational choice theory—to help understand and predict how government policies develop. In some cases, fear of subsequent liability, whether warranted or not, inhibits adoption of a mitigation option, particularly on the part of local governments.

One would hope that federal, state, and local policymakers' processing of information regarding hazards would be systematic rather than heuristic or otherwise flawed. If so, information programs for them could emphasize scientific information and attribute portrayal—models of estimated losses and information about alternative mitigation measures. But the degree to which such tools would increase local policymakers' adoption of hazard mitigation remains to be determined. No studies indicate whether state and local policymakers perceive those who design, develop, and operate loss estimation models as credible; whether their past experience inclines them to have confidence in computer models; or whether the models would provide all of the information they need to make a decision in favor of adopting mitigation measures.

The willingness or reluctance of policymakers to require local efforts to avert losses is shaped by a variety of forces. State and federal governments have a strong stake in encouraging local governments to reduce potential losses, if only to lessen state and federal disaster relief. Many studies point to the reluctance of local governments to adopt on their own, or to adequately enforce, strong measures for managing land use and development in hazardous areas. State requirements for local hazard planning vary according to the nature of the hazards, differing perceptions of the seriousness of hazards, and differing beliefs about appropriate state intervention. Studies of state-level attitudes about hazard mitigation show that influential groups identified by state policymakers do not see it as an important issue. National policymakers face pressure to reduce federal disaster relief outlays. However, they are also cognizant of strong concern among state and local officials about "unfunded mandates" along with the political power of the property rights movement. Little is known about the extent to which government agencies attempt to minimize their own losses and to assure continued provision of routine government services after a disaster.

SOCIAL INFLUENCES

Besides the internal processes that an individual or group uses to make decisions about mitigation, there are a variety of other factors that may influence the decisionmaking process itself or may influence the outcome in other ways.

The Nation's Culture

In the United States and much of the western world there is a belief that technology can and eventually will make humans safe from all of the forces of nature. Western peoples tend to believe and act as if losses are caused by surprise extreme events, rather than by choices about how and where buildings are located and other development takes place. There is a widespread self-interest motivation and a majority-rules mindset.

Individualism and the sanctity of private property are important cultural values in the United States, and they influence the laissez-faire, persuasion-oriented approach that is generally taken to encourage hazard reduction activities. While in the interest of public safety the government does exercise some control over what individuals do, that control does not extend to requiring mitigation efforts that could reduce disaster losses. This same respect for property and the right to accumulate profits underlies resistance to mitigation measures that might place additional "mandates" on business owners.

For example, the contemporary consensus in the United States is that warnings provide advice and recommendations but do not mandate action. Sometimes this has meant standing by as some people decide not to evacuate and face almost certain death: for example, officials at Mount St. Helens knew that some residents were refusing to leave. In the same vein the value placed on individualism also shapes the selection of loss reduction strategies. In attempting to enhance preparedness and appropriate response behavior in the United States, the most common approach is to focus on individual households. Less emphasis is placed on neighborhoods or other groups. Despite the American focus on individualism, though, altruism is also a very strong cultural force. Throughout its history the United States has had a strong tradition of volunteer behavior and community involvement, and this altruistic orientation carries over and is amplified in disaster situations. Cultural expectations also prepare specific occupational groups, such as firefighters, police, paramedics, and other emergency workers, to put their individual self-interests aside.

Thought Processes

Today's complex world is characterized by growing sets of interrelated problems, and change can occur quickly and in unexpected ways. Unfortunately, efforts at managing our world use a traditional planning model: study the problem, develop alternatives, choose one, and move on to the next problem. This is based on the view that more applications of existing knowledge will result in proportional advances in social adjustment and hazard reduction. This perspective assumes that hazards are fairly static and sees hazards mitigation as an upward linear trend in the sense that any mitigation is desirable. Thus, linear, unintegrated, static mitigation solutions are being proposed to cope with complex, interactive, nonlinear hazards problems. In addition, most individual and collective decisionmaking is characterized by short-term thinking.

Institutional Factors

Society today is not organized to allow hazards to be viewed in nonlinear ways or to share knowledge and experts across disciplines or practitioners across different levels of government. The current management structure is based in specialized problem-solving departments. A consequence is that each division fails to recognize the integrated character of problems—especially environmental ones, including natural and related technological hazards. For example, hazards mitigators, emergency planners, resource managers, and community planners all seek to solve hazards problems on their own: such isolated efforts are inadequate to an interdisciplinary task. An analogous situation exists in the educational sector, where university-level education in the fields that could support sustainable hazards mitigation is strictly segmented, and there is little encouragement for interdisciplinary work.

One example of how the institutional structure hinders comprehensive mitigation approaches is in the overlay of various federal and state mandates that attempt to influence local decisions about land use and development in hazard-prone areas. The result has been a patchwork system arising from changes in legislation over time, variation in requirements across hazards, and differences in demands from one agency to another. In addition, this governance is highly fragmented and resistant to change. Many local officials perceive federal and state environmental mandates as overly prescriptive and coercive. Local governments complain about the failure of higher-level governments to fund the costs of implementation, the lack of flexibility in the required actions, and the shifting to them of political blame for infringement on property rights. For their part, local governments mismanage hazards-related initiatives and are only too happy to reap the benefits of federal programs.

Cultural Influences

In the past 20 years there has been a growing recognition of the ways in which a range of sociocultural and sociodemographic factors influence hazard mitigation, disaster preparedness, and response.

Culture

There are many ways in which cultural influences affect people's actions with regard to hazards and disasters. Behavior influenced by cultural or ethnic differences cannot always be generalized to other people. Only a limited amount of systematic research has examined the degree to which cultures differ in their perceptions of the characteristics of hazards and adjustments. For example, it is not clear whether all people are susceptible to the kinds of errors in risk perception and decisionmaking that have been documented in research on Americans.

Cultural theories also suggest that the ways societies organize themselves to address different hazards help focus attention on some hazards, such as nuclear power, while downplaying others, such as earthquakes. They also contend that the cultural context may increase or reduce awareness of hazards, constrain individual decisionmaking by facilitating some alternatives while precluding others, and alter the range of acceptable responses. Repeated instances in the United States indicate that a community's response to scientific information about natural hazards can be significantly affected by the boosterism frequently found in local culture.

Race, Ethnicity, and Gender

A body of work is accumulating that demonstrates that racial, ethnic, and gender differences influence a wide range of perceptions and behaviors, including threat perception, concern about a hazard, understanding of and belief in the science underlying hazard information, attitudes toward the entities disseminating information, response to warnings, shelter-seeking behavior, perceptions of the credibility of authorities, and making decisions about evacuating or taking other mitigative action.

As an example, language can constitute a barrier and influence behavior in various ways, from making the receipt of warnings or hazard-related information problematic to limiting access to information on options for shelter and other services.

Social Bonds

Social attachments and relationships help determine the preparedness, mitigation, and response behaviors undertaken by different people. Friends, relatives, neighbors, co-workers, and the news media influence the adoption and implementation of hazard adjustments. Social bonds have been shown to foster adaptive behavior both before and after disasters. Links and ties to others are important for organizational preparedness and response, too. Interorganizational linkages are major sources of information transfer and of access to new ideas and practices.

Informal social influences on hazard adjustment are most prevalent when people do not have an opportunity to learn directly from their physical environment. When the physical environment is too complex or rapidly changing, people tend to be influenced more by other people than by the physical reality of the hazard itself.

ECONOMIC FACTORS

Although it is known that humans do not make decisions according to the classic economic model of weighing costs and benefits, economic considerations are nevertheless among the most powerful influences on mitigation.

Constraints

It is known that households and businesses tend to favor easy and inexpensive mitigation measures, rather than those that will be most effective in reducing damage. A positive relationship generally exists between household income and adoption of preparedness measures. Relatively inexpensive and easy preparedness strategies, such as having a battery-operated radio and a working flashlight and storing water, are far more likely to be used than more expensive or time-consuming ones. It is also likely but not documented, that financial resources are a significant indicator of preparedness at the organizational and interorganizational levels. The mix of preparedness strategies adopted by communities is also strongly influenced by economic considerations. It has been argued that simpler preparedness and response activities are emphasized over longer-term mitigation in most U.S. communities precisely because they are less expensive and easier to sell politically.

The reluctance of many communities to adopt warning systems provides a clear illustration of this. Such systems are often a high-cost item with no immediate payback. They also compete against other more visible public safety expenditures such as conventional police and fire services. If investments in warning systems were made on the basis of conventional cost-benefit analyses, few warning systems would be adopted. Instead, warning systems often are approved based on public humanitarian sentiments after a particular emergency, regardless of the outcome of a cost-benefit analysis.

Incentives

As it functions today, U.S. policy in general permits vulnerability to escalate while failing to provide economic incentives for risk avoidance. In a system that does not provide tax relief or other incentives, decisions by homeowners and businesses to take only the least expensive mitigation options (rather than those that will best reduce future damage) are economically rational.

Economic incentives to encourage mitigation could take the form of subsidies, low-interest loans for retrofitting, or tax breaks or relief for mitigation activities. But incentives have advantages and disadvantages. On the one hand, they may be effective in inducing adjustments by lowering the cost of implementation. But to the extent that they subsidize vulnerable populations that should be protecting themselves, they may be subject to criticism. The net effect of well-designed incentives would be that individuals and organizations in hazard-prone areas reap benefits by reducing their vulnerability, but society as a whole benefits because less tax money is spent on subsequent disaster assistance and recovery.

LEGAL CONSIDERATIONS

Mandates

One strategy for inducing adoption and implementation of hazard mitigation, primarily used by governments, is simply to require adoption. Building codes are among the most successful natural hazard mitigation mandates. Merely passing laws and issuing directives do not ensure that desired changes will actually take place, but there is evidence that mandates and other types of legal and regulatory requirements can have a positive impact under some circumstances and that the absence (or poor design) of such requirements can slow hazard reduction efforts.

For example, it is unlikely that formal disaster plans would have become almost universal at the local level if they had not been required. Similarly, it is clear that changes in the regulatory environment after the Three Mile Island accident did spur nuclear facilities and nearby communities to expand their planning efforts. Without the Superfund Amendments and Reauthorization Act (SARA) Title III, it is unlikely that local communities with hazardous materials facilities would have developed multiorganizational networks to deal with emergency planning and response for chemical hazards.

Likewise, most of the impetus for adoption of community warning systems, for example, comes from a need to protect public safety and the political and legal consequences to local governments of failing to do so (see discussion on legal pressures below). One noteable exception may be the early earthquake warning system being adopted in southern California with little feasibility or cost-benefit study.

At the same time, it is evident that not all mandated programs achieve their objectives. Some generate resistance rather than compliance, as illustrated by the Federal Emergency Management Agency's ill-fated 1980s crisis relocation program. Unfunded state and federal mandates on local governments are politically unpopular locally and can cause resentment. Sometimes it is difficult to get hazard mitigation mandates enacted. Mandates that are too simplistic force some individuals and organizations into excessive levels of protection, while others escape without enough. Unfortunately, mandates that are complex enough to avoid these problems frequently cannot be understood by nonspecialists.

Another problem is that mandates can ignore people's ability and motivation to comply. These impediments can arise in households, as well as in private- and public-sector organizations. Building inspectors are an excellent case in point. Whatever the state or other requirement, a jurisdiction must have a desire to properly enforce its building codes. Likewise, however motivated they may be, building officials' ability to do the job satisfactorily depends on budget support, personnel recruitment and selection, equipment, other local laws, job design, and job training.

Furthermore, sanctions that appear reasonable can be defeated in a variety of ways. For example, it has been documented that requirements for disclosure of natural hazards before the sale of real estate have been circumvented by burying the disclosure in a mass of paperwork, providing it in cryptic terms, or delaying the disclosure until the final papers are ready to be signed.

Legal Pressures for Adoption and Implementation

The courts have been quite clear that local governments have an obligation to adopt programs and practices to protect public health and safety, and this can include hazard mitigation. Courts have granted relief to property owners who have suffered losses because of the failure of a governmental entity to enact or implement such programs, as in the case of failure to maintain canals or other drainage systems. The courts have been more divided over the financial responsibility of local governments if they fail to plan effectively, implement land-use plans, enforce building codes, or provide adequate emergency response. This situation influences the propensity of local entities to adopt, enforce, and implement at least certain mitigation measures.

On the other hand, sometimes fear of subsequent liability, whether warranted or not, inhibits adoption of an adjustment, particularly on the part of local governments. Many hazards managers perceive a great risk today of civil claims and court judgments. For emergency planning, response, and recovery efforts, the application of tort law is quite clear: state courts have consistently granted immunity to public, private, and nonprofit actors involved in those activities. But the law is less well defined for efforts to mitigate anticipated damage.

For many years, states and political subdivisions enjoyed immunity under tort law from lawsuits that sought relief for the wrongful conduct of governmental officials or employees. But by 1995 a series of state court decisions and legislative actions had eroded somewhat the principle of sovereign immunity, so that in many states there is some latitude for claims to be made against public agencies. In the absence of immunity, government agencies are most often sued for damages for alleged negligence in the performance of their duties—for example, a failure to conduct an inspection, as required by statute.

Discretionary immunity grants immunity for the judgmental decisionmaking process of public officials and employees and is intended to free officials from fear of liability if that judgment results in harm to another. For example, the decision whether to deploy emergency response units in a particular fashion is discretionary. The decision to adopt a local ordinance or public program is a high-level policy decision and likely to be viewed by the courts as a discretionary action.

In a few instances, communities and states have attempted to enforce land-use ordinances, but the courts have sided with citizens on the issue of rights to develop one's own property as long as adequate solutions—always engineered ones—are incorporated. Decisions like this have caused a "chill" in legislative attempts to use land use for mitigation purposes. The duties and potential liability of governmental units are shaped by many factors, including immunity for specific functions of a jurisdiction, the discretionary functions of government officials, and immunity provisions in state statutes.

OTHER INFLUENCES

There are at least two other influences on adoption and implementation that bear mention. The first is the noneconomic incentive, and the second is the public awareness program—actually an amalgamation of techniques geared toward influencing people's perceptions and behavior.

Noneconomic Incentives

Noneconomic incentives have been offered, with apparent success, to encourage people to heed warnings and evacuation notices. They have included information hotlines, transportation assistance, mass care facilities, and security and property protection. Federal and state governments encourage hazard mitigation at lower levels by offering the incentives of technical assistance, reduced administrative oversight, and limited regulation—all of which usually are attractive to local governments.

Public Awareness Programs

There are numerous public and private hazard awareness programs throughout the United States, but few studies have been conducted on whether they follow the principles of effective communication or are successful in promoting hazard mitigation. The few that have been done were based on only a small set of awareness programs and have reported inconsistent results. Despite their limitations, it is possible to draw some conclusions from the available studies of hazard awareness programs by supplementing them with information from cross-sectional studies of the adoption and implementation of hazard adjustments.

Information Sources and Channels

Information sources for hazard awareness programs have been categorized as authorities (federal, state, and local government), news media (print and broadcast), and peers (friends, relatives, and neighbors). These sources differ in their ability to influence an individual's willingness to attend to and accept hazard information. Official sources are the most credible. Existing research seems to indicate that hazard awareness programs increase in effectiveness when they rely on multiple sources who repeatedly transmit a variety of messages through many different outlets.

Messages disseminated over different media inherently have different characteristics. So different channels probably contribute to different stages of information processing. For example, radio or television spots may be best at initiating or maintaining hazard awareness, while printed materials may be best at providing detailed information about mitigation.

Message Characteristics

Messages have several characteristics: amount of material, speed of presentation, number of arguments, repetition, style, clarity, ordering, forcefulness, specificity, consistency, accuracy, and extremity of the position advocated. Some of these characteristics can be measured objectively; others are more ambiguous.

Hazard awareness programs can also be classified by their different themes. Some programs are designed to attract attention, others to enhance acceptance of the message, using mascots like Owlie Skywarn or celebrity endorsements.

Other programs focus on content, such as scientific information programs that disseminate technical data about a hazard. Such information is processed and retained by only a relatively small number people. In contrast, practical instructions focus more on protective response than on the hazard itself. The very simplest form of practical instruction is the prompt, a sign that defines a single contingency ("Climb canyon wall in case of flash flood"). Prompts are more likely to attract attention, be readily comprehended, and retained for future use. Other message styles are attribute portrayal strategies, which emphasize the advantages of recommended hazard adjustments, and fear appeals, which attract attention and motivate action by describing the potential personal consequences of disaster.

Receiver Characteristics

Most hazard awareness and education programs have assumed a homogeneous "public" and done little to tailor information materials to different groups. One obvious reason is that it is cheaper and easier. But many researchers have emphasized the importance of tailoring information to the characteristics of the audience. This advice has been implemented by a few hazard awareness programs, but there is a conspicuous lack of definitive guidance on the topic. The Bay Area Regional Earthquake Preparedness Program, the Southern California Earthquake Preparedness Program of the California Office of Emergency Services, and the American Red Cross have published guides and manuals for special groups, including schools, hospitals, corporations, city managers, emergency managers, and the media, but the success of these targeted approaches has not been studied.

The fact that different communications programs focus on different components of persuasion as well as different phases of information processing suggests that these themes are neither mutually exclusive nor exhaustive. Thus, it may be possible to create more effective hazard awareness programs and influence people to take mitigative action by presenting different themes in successive messages. Unfortunately, no attempts have been made thus far to use these principles to design a hazard awareness program.

CONCLUSION

Much of the knowledge in hand today about the adoption and implementation of hazards mitigation and preparedness measures comes from research on discrete individual mitigation actions based on the linear model of problem solving and behavior. A looming question is whether the knowledge summarized in this chapter is applicable to sustainable hazards mitigation, since it requires a broader perspective and nonlinear approaches. Future research on the adoption and implementation of sustainable hazards mitigation need not exclude important topics suggested by the knowledge in hand, but it will require an additional knowledge base generated from a different line of inquiry. That approach would be broader in subject matter and would root adoption and implementation in factors that have received little hazards-specific research attention so far—public sentiment and support for conceptual and cultural shifts, changes in attitudes, and alterations in how organizational and governmental mitigation approaches are structured—as much as individual decisions and behavior. In addition, it would be wise to target future adoption and implementation research on those specific mitigation activities that will truly build sustainability.

CHAPTER SIX

Tools for Sustainable Hazards Mitigation

Of the many techniques for coping with natural hazards and disasters that people have attempted to use, the five categories summarized in this chapter have proven over the past two decades to be the most useful for minimizing and/or redistributing losses and reducing social and economic disruption. Each of these mitigation tools (or adjustments as they were labeled in the first assessment)—land-use planning, building codes, insurance, engineering, and warnings—is supported by its own body of research, disciplines, experts, and government and private-sector management structure. These five mitigation tools are reviewed, and knowledge about them is summarized in this chapter. Also discussed are their potential roles and contributions to sustainable hazards mitigation. Absent from this chapter are discussions of disaster preparedness, response, and recovery, which are addressed in Chapter 7.

LAND-USE PLANNING AND MANAGEMENT

No single approach to bringing sustainable hazards mitigation into existence shows more promise at this time than increased use of sound and equitable land-use management. Many political, social, and economic forces conspire to promote development and redevelopment patterns that set the stage for future catastrophes. However, by planning for and managing land use to accomplish sustainable hazards mitigation, disasters—though not wholly eliminated—can be reduced to a scale that can be borne by the governments, communities, individuals, and businesses exposed to them.

Land-use planning, environmental protection, hazards mitigation, and sustainable communities are integrally related concepts with a similar vision—communities where people and property are kept out of harm's way from natural disasters, where the mitigative qualities of natural environmental systems are maintained, and where development is designed to be resilient in the face of natural forces. The landscapes of sustainable communities incorporate compact, higher-density development and more efficient use of land and space; greater emphasis on trees, parks, and open space; redevelopment of underutilized urban areas and infill development; public transit; mixed-use environments that are more amenable to walking and less dependent on automobiles; energy and resource conservation; renewable energy; and minimizing waste and pollution. Increasingly, community resiliency to natural disasters is being added to this list (see Berke, 1995).

Comprehensive Local Land-Use Plans

Local governments have many land-use management tools at their disposal for averting disaster losses and increasing community sustainability: building standards, development regulations, critical and public facilities policies, land and property acquisition, taxation and fiscal policies, planning processes, and information dissemination. Communities must decide which combination of measures will be effective, efficient, equitable, and feasible for them. An integrated, comprehensive community plan ties hazards mitigation, land use, and environmental, social, and economic interests together and lays out guidelines for when and how these tools are to be used. With the right mix of land-use management measures, local governments can reduce disaster losses while accomplishing environmental and other community goals.

Recent studies in Australia, New Zealand, and the United States have documented a number of benefits that follow when governments plan before they act. First, by providing information about the location and nature of various hazards, plans ensure that the limitations of hazard-prone areas are understood by policymakers, potential investors, and community residents. Second, by indicating the most appropriate uses of land in a community (and showing that hazardous areas do not always have to be used more intensively for communities to realize economic and other development objectives), plans make it possible for communities to consider and, where economically efficient, actually adopt restrictions on building in hazardous areas. Third, good land-use plans help educate the public, and this education, in turn, increases demand for action. Also, education used in concert with regulation is likely to be more effective than regulation alone (see Burby and May, 1997; May et al., 1996).

With a long-range, comprehensive, sustainability-oriented plan, a community can coordinate multiple issues, goals, and policies effectively. Local land-use plans should (1) be made before disasters occur (although better late than never); (2) be oriented toward the long-term future; (3) be focused on systems, as well as parts (ecosystems and cumulative effects vs. small areas and individual disasters); (4) include all sectors of society in the decisionmaking process; (5) begin by assessing hazards and characteristics of the community; (6) determine values, goals, and objectives; (7) adopt unambiguous plans and policies; (8) have provisions to monitor and evaluate the effectiveness of the policy; and (9) be flexible enough to change with time. As an illustration, among the sustainable mitigation components of a good comprehensive plan would be the following:

- *Hazard identification*: magnitude, location, and probability of a disaster.
- *Impact assessment*: what populations and properties are exposed to hazards, and the likely damage in a disaster.
- *Loss estimation*: the quantitative probability of damage, injuries, and cost in a given area over a specified time.
- *Carrying-capacity assessment*: the maximum load (population × per-capita impact) that can safely and persistently be imposed on the local environment by society without reducing the ability of the environment to support such a community in the future.
- *Built-out analysis*: the maximum level for the buildings and infrastructure given the character of the local social and environmental systems.
- *Ecological footprint analysis*: an estimate of the land and water area needed to support local consumption and development practices.

- *Assessment of sustainability indicators*: Many communities (e.g., Boulder, Seattle, Chattanooga, Tallahassee) have identified and measured such indicators as education, the economy, public safety, the natural environment, health, the social environment, politics, culture, and mobility.
- *Environmental impact statement*: Such a statement should always include an analysis of natural hazards.

Federal and State Policies

If local governments are to prepare and follow through with land-use planning for hazard mitigation, experience suggests that nothing short of strong mandates backed by commitment and effort by the federal government and states will suffice to bring it about. Without them most local governments will continue with business as usual. A few innovative jurisdictions—those with extraordinary local leadership and those that have suffered severe losses in the past—will plan for and manage land use in hazardous areas. However, most will not. Hazard mitigation requires a partnership. The actual planning and conduct of programs must occur at the local level, but a great deal of the impetus must come from above.

At present there is no overarching federal policy that governs land use and development in hazard-prone areas. Instead of a holistic approach, there are pieces, including over 50 federal laws and executive orders that relate to hazards management. Federal and state governments seem unwilling to embrace direct land-use management, yet they send local governments conflicting signals about exposure to hazards. Policies advocate risk reduction and transfer (standards, insurance, relief, etc.) rather than risk assumption and elimination. While such policies are appropriate in some cases, they also have created problems. Uniform standards disregard the fact that in some locales it is appropriate to assume the risk of developing hazardous areas. The focus of and subsidies built into federal insurance and relief programs probably help account for the massive increase in development in hazard-prone regions (see Burby and French, 1985; Beatley et al., 1994; Platt, 1987). The assurance of federal assistance after a repeat disaster creates a "moral hazard"[1] by lowering the incentive to avoid risk. Finally, federal policies

[1]The term "moral hazard" is used to describe a situation in which one of the parties to an agreement has an incentive, after the agreement is made, to act in a manner that benefits

arguably have lessened the chances that hazard-prone development will be exposed to short-term losses (e.g., from a 100-year storm), while allowing the potential for greater losses from disasters with longer return frequencies to grow.

The federal and state focus on risk reduction and risk transfer, besides increasing exposure to hazard, has effectively shifted liability for the occupation of hazardous areas to Washington and to a lesser degree to the state capitols, thus relieving local governments of their traditional responsibility for managing these areas. The federal and state governments' top-down approach to dealing with local stakeholders has done little to foster the "local involvement, responsibility, and accountability" called for in the most recent comprehensive review of federal policy (Interagency Floodplain Management Review Committee, 1994, p. 82).

Also, because governmental institutions are disjointed and disaster policies are not integrated into general land-use policies, a host of other nondisaster-related federal policies limit the ability of local governments to use land-use management to reduce exposure to hazards. For example, increased exposure of people and property to flooding has been abetted by federal financing of highway construction, sewers, and other infrastructure that increase development potential in flood-prone areas while also reducing development costs. Some progress has been achieved in integrating hazards considerations into local planning and land-use management, however, particularly in states that have comprehensive growth management programs.

Barriers to Local Action

In addition to hindrances and disincentives inherent in state and federal policies, there are many other reasons for local government reluctance to adopt strict land-use policies. The most important are a lack of local political will to manage land use, deficiencies in management capacity, and regional fragmentation.

---

him- or herself at the expense of the other party. Moral hazard exists in the case of insurance when an insured person has no incentive to avoid the a risk and, in fact, may take action to bring it on in order to collect a financial payoff. It also affects government programs that provide benefits (such as an enhanced level of protection from a hazard such as a flood), thereby relieving the people who benefit from the protection from having to take responsibility for mitigation.

Crisis of Commitment

Few local governments are willing to reduce natural hazards by managing development. It is not so much that they oppose land-use measures (although some do), but rather that, like individuals, they tend to view natural hazards as a minor problem that can take a back seat to more pressing local concerns such as unemployment, crime, housing, and education. Also, the costs of mitigation are immediate while the benefits are uncertain, do not occur during the tenure of current elected officials, and are not visible (like roads or a new library). In addition, property rights lobbies are growing stronger. All of these factors contribute to a lack of political leadership for limiting land use in hazardous areas.

Shortfalls in Capacity

The science of identifying hazards and designing to reduce their adverse impacts has far outrun the ability of local governments to put this new knowledge into practice. Hazard-zone mapping is enormously expensive. Few planning programs provide detailed instruction in hazard mitigation. Many enforcement personnel have insufficient knowledge or support to enforce hazards-related code provisions effectively. Disagreement among state and local policymakers, lack of political power to influence development and environmental interests, and few perceived alternatives all contribute to a lack of capacity on the part of local governments.

Failure to Act Regionally

Yet another problem stems from the fact that hazards do not respect political boundaries and vice versa. In some cases, land-use management programs for hazard mitigation cannot be effective without cooperative intergovernmental coordination, which is difficult to achieve. Such fragmentation also tends to work against considering the cumulative effects of land-use changes.

Conclusion

Promoting appropriate land use has the great potential of not only keeping people and property out of harm's way but also of providing more affordable housing and living conditions, protecting important environmental resources and amenities, and reducing the long-term costs of growth and development, among others. Comprehensive, locally based, land-use management can go a long way toward building sustainable communities.

BUILDING CODES AND STANDARDS

The quality of buildings and other structures plays a direct role in determining lives lost, injuries, and the financial costs of disasters. So disaster-resistant construction is an essential component of local resiliency to disasters. Sustainable hazards mitigation requires that engineered solutions be used wisely and in balance with other approaches to enhance resiliency, environmental quality, economic vitality, intra- and intergenerational equity, and quality of life. The use of building codes can strengthen the nation's constructed environment.

Codes and Standards

The regulation of building construction in the United States is accomplished through building codes. A building code is a collection of laws, regulations, ordinances, or other statutory requirements adopted by a government legislative authority having to do with the physical structure of buildings. The purpose of a building code is to establish the minimum acceptable requirements necessary for preserving the public health, safety, and welfare as well as the protection of property in the built environment. These minimum requirements are based on natural scientific laws, on properties of materials, and on the inherent hazards of climate, geology, and use of a structure.

The primary application of a building code is to regulate new or proposed construction. It has little application to existing buildings unless they are undergoing reconstruction, rehabilitation, or alteration or if the occupancy category is being changed. The term "building code" is frequently used to refer to a set of code books that are coordinated with each other to address specific technical applications. This set of codes generally consists of four documents: a building code, a plumbing code, a mechanical code, and an electrical code. This division is more for convenience than for specific technical or legal reasons.

A standard is a prescribed set of rules, conditions, or requirements with definition of terms; classification of components; delineation of procedures; specification of dimensions, materials, performance, design, or

This home in the Pacific Palisades section of Los Angeles was destroyed in the 1994 Northridge earthquake. Photograph courtesy of AP/Wide World Photos.

operations; description of fit or measurement of size; measurement of quality and quantity in describing materials, products, and systems; and services or practices. There are hundreds of standards in the United States addressing virtually every construction applications, from design practices and test methods to material specification. The building code coordinates this massive quantity of information into an orderly, intelligible, responsive system to safeguard health, safety, general welfare, and property.

There are three basic classifications of standards used in building codes. They are engineering practice standards, materials standards, and test standards. Engineering practice standards define methods of design, fabrication, or construction and specify accepted design procedures, engineering formulas and calculation methods, and good practice. Materials standards are specifications establishing quality requirements and physical properties of materials or manufactured products. Test standards include structural unit and system tests, durability tests, and fire tests. When a particular property of a material, product, or system is required by the building code, the code will specify the standard to which the product is to comply or be tested. It will also state the criteria for determining code compliance.

Model Building Codes

There are three model code organizations active in the United States today. The first, Building Officials and Code Administrators International, Inc. (BOCA), was created in 1915, and represents the eastern and midwestern portions of the United States. The second is the International Conference of Building Officials (ICBO), which produces codes for the West and Midwest. The third, the Southern Building Code Congress International, Inc. (SBCCI), represents the South and Southeast. BOCA publishes the National Building Code, ICBO writes the Uniform Building Code, and the Standard Building Code is produced by SBCCI. Each of these organizations is a nonprofit public benefit service corporation owned and governed by its voting members, which are units of city, county, and state governments as well as the federal government. Each organization develops model codes, provides training in all aspects of codes and code enforcement, and conducts other activities that benefit its members.

In 1994 the International Code Council (ICC) was established to develop a single set of comprehensive and coordinated construction codes. The ICC Board of Directors consists of three representatives from each of the model code organizations. The new codes are referred to as International Codes and are being developed using criteria based on the three existing model codes. A model code has no legal standing until it is adopted as law by a state or local jurisdiction. All owners of property within that jurisdiction are then required to comply with the enacted building code for improvements to their property. The model codes are used as references by design professionals even where the codes have not been adopted in a specific area.

Local and State Codes

Local and state codes vary considerably in degree and procedures. Practically all are based on one of the model codes. At the time of the first assessment (White and Haas, 1975) two decades ago, very few states had state building codes; codes were locally enacted. Since then about half of the states have retracted this delegation of power to the local government and have enacted a state code. The state building code preempts the local government's authority to enact a local code with the same scope and application. The state legislatures have generally taken this action for two reasons: to provide equal protection to all citizens throughout the state and to develop statewide uniformity for commerce purposes.

Local governments have traditionally enacted comprehensive building codes that regulate all construction. The current trend is for states to increase the application of their statewide building codes by replacing the laws that had limited applications. Additionally, state and local governments are relying less on their own custom-drafted building codes and are adopting model building codes, thereby diminishing complexity.

State codes vary from those that merely adopt a particular edition of a model code with administration left to local jurisdictions to those that start with a model code and revise and administer through a separate state-established code body. Many states adopt and administer a separate code only for state-funded buildings, while others may require a special code for certain occupancies, such as schools and assembly buildings.

Local codes are also diverse. Most local amendments are limited to administrative provisions, which are subject to change to meet other local regulations regarding implementation or ordinances.

Enforcement

Even with a statewide code, the administration and enforcement of all building codes rests with the local governments, with varying degrees of state oversight. The local government is responsible for creating the organizational structure for the code enforcement process, designating the person or persons responsible for enforcement, and providing the necessary resources. The local enforcement entity, directed by the code enforcement official, can come in any size or shape. It could consist simply of a one-person department that reports directly to political leaders, or it could be a larger organization that has specialists in all engineering disciplines and operates as a major city or county department. The size and shape of each organization are determined by the amount and nature of construction activity, the relative importance of code enforcement in the priorities of the jurisdiction, and the resources—especially financial—that are available to support the activity.

A building permit process is established for the review, inspection, and approval of proposed activities to secure compliance with the building code. A certificate of occupancy is issued when all inspections have been performed, any deficiencies corrected, and the construction work completed. Any code is only as good as the enforcement that goes along with it. For example, south Florida was thought by most people to have the most rigorous building code in the country, although even it had some deficiencies. However, Hurricane Andrew proved that a rigorous code is ineffective if it is not properly enforced (see Chapter 4).

A survey of building department administrators from jurisdictions of all sizes in the southeastern United States was performed in 1995. Its goal was to assess the perceptions of building code professionals regarding the adequacy of resources at their disposal to administer and enforce codes in their jurisdictions. About half reported that their departments were not adequately staffed to perform all necessary inspections or handle all plan review responsibilities (see Insurance Research Council and Insurance Institute for Property Loss Reduction, 1995).

In an effort to obtain better code enforcement, the Insurance Institute for Property Loss Reduction (now the Institute for Business and Home Safety) instituted the Building Code Effectiveness Grading Schedule. Eventually, every building department in the United States will be evaluated and assigned a rating, which can affect the personal line insurance rates of the community (analogous to the one communities already use for fire protection). It is anticipated that this activity will go a long way toward improving and maintaining rigorous building code enforcement.

Currently, however, there are signficant reasons why building codes may fall short of their loss reduction potential. For example, building codes are for life safety and do not provide for property protection or functionality after a disaster; many local jurisdictions do not have a building official or department; many states allow local jurisdictions to petition waivers from the state-required building code; and state-mandated codes are often reserved only for certain types of buildings and not for most commercial or residential structures.

INSURANCE

Most property owners rely on private insurance to protect themselves against financial losses from ordinary natural disasters such as fires and windstorms. Insurers generally are able to pay for these common losses from the annual premiums collected, supplemented by investment income and reinsurance recoveries. However, most property owners do not currently buy coverage against special perils, notably earthquakes, hurricanes, and floods.

Insurance is now available for some but not all natural disaster agents. It varies from state to state and among carriers. Insurance coverage is nearly universally available for wildfires, winter storms, volcanoes, tornadoes, lightning, and hail. These perils are covered under most standard property insurance contracts. Generally speaking, these events are sufficiently random and widespread to permit the private insurance mechanism to operate effectively.

Landslides are normally not considered an insurable peril by private insurers. They are covered by insurance programs only if the damage is caused by an insured earthquake or flood damage covered under the National Flood Insurance Program (NFIP). Hurricane wind damage is included as part of the basic wind coverage in most property insurance policies. Flood damage from hurricanes is not included but can be purchased separately under the NFIP.

Insurance coverage for damage from earthquakes is not automatically included in homeowners' insurance policies, but it can be purchased as a rider for an additional premium. Earthquake coverage is often included on commercial policies for structures in hazard-prone areas, and it can also be purchased separately. Protection against loss by fires that might follow an earthquake is included in the basic fire peril coverage in all property insurance contracts; business interruption from an earthquake is covered by a separate policy. Until the 1980s fewer than 10 percent of property owners in California had purchased earthquake insurance. Even in areas of relatively high earthquake loss probability, only between 30 percent and 40 percent of property owners purchase the coverage today. Recent earthquakes coupled with a legislative requirement that insurers offer earthquake coverage on a biennial basis have increased the number of California property owners who purchase earthquake insurance.

Insurance companies have viewed flood risk as uninsurable because of problems of both adverse selection and highly correlated risks. In 1968 Congress enacted the NFIP as a means of offering coverage nationwide through the cooperation of the federal government and the private insurance industry. Under this program the federal government conducts the hydrological studies needed to identify areas vulnerable to floods and to define the nature and extent of the hazard. State and local governments are responsible for adopting and enforcing minimum standards for floodplain zoning and construction. Private insurers sell and service the federally underwritten insurance policies, subject to federal standards.

The insurance component of the NFIP is provided at lower rates (subsidized by the premiums paid by other policyholders) for older structures that existed in flood hazard areas before the program began. The number of properties qualifying for subsidized rates has declined over time, while the rates charged have gradually increased to make the program as a whole self-supporting in years with average flood losses. The NFIP borrows from the U.S. Treasury in years when catastrophic losses exceed the amount accumulated in the insurance fund.

Who Should Pay for Disaster Losses?

To address the appropriate role of insurance in sustainable hazards mitigation, the following question needs to be posed: Who should pay for disaster losses? A society that believes that every citizen should share in the losses of disaster victims may find that taxation is the best way to provide the revenue to cover these costs. If, on the other hand, society believes that individuals are responsible for bearing their own burdens, some form of insurance with risk-based rates may be appropriate.

This brings up the question of whether certain individuals or groups should get special treatment at the expense of others. If so, the treatment should be such that situations will not be created that have long-term negative consequences. For example, if uninsured disaster victims are guaranteed grants and low-interest loans that enable them to continue to locate their properties in hazard-prone areas and more people continue to move into the hazardous areas, taxpayers will be subject to increasingly larger expenditures for bailing out more victims in the future. What may be viewed as equitable immediately after a disaster will be seen as inefficient from a long-term sustainability perspective.

If certain victims' disaster costs are to be subsidized by others, private risk-based insurance cannot be counted on to cover damage from these events over a long period, although it can be prevailed upon for a limited time. In the long run such socially motivated subsidies can only be successfully achieved and maintained by some form of government insurance. Historically, attempts to require private businesses to overcharge some groups in order to subsidize others have broken down in a competitive marketplace, despite more and more elaborate enforcement procedures. There is an important difference in the underlying principles of private insurance, where premiums are based on risk and a system in which taxpayers are expected to absorb disaster costs for that segment of the population deemed to require special consideration.

The Demand for Disaster Insurance

Many residents of floodplains and earthquake fault zones choose not to insure their homes and businesses against flood and earthquake damage, even though about 95 percent of them buy insurance for fires, wind storms, and other common perils (see Insurance Research Council, 1996). Only about 20 percent of the homes exposed to flooding are insured against floods, and fewer than that are insured against earthquakes countrywide. However, the demand for earthquake coverage is rising, especially in California. Flood and earthquake coverage of business property also is less than 50 percent (Insurance Research Council, 1991). The result is that millions of property owners in high-risk areas can be expected to turn to federal disaster relief programs in the event of catastrophic floods and earthquakes, placing a heavy burden on taxpayers.

The question raised by such facts is: What factors affect the decision to purchase, or not purchase, disaster insurance? Despite the rational and even highly mathematical calculations of human behavior by economic theorists, the plain fact is that few individuals behave according to economic theory. People do not rationally weigh the costs and benefits of various strategies and then select the one that minimizes costs and maximizes benefits (see Chapter 5).

Researchers who have studied the low demand for flood and earthquake insurance suggest several reasons why many people do not buy these optional coverages. In general, they are much the same as the influences on adoption and implementation of mitigation measures discussed in Chapter 5. In short, people think the premium is too great an expense for an uncertain payoff possibly far in the future; they think that it can't happen to them; they think that federal assistance will make them whole if a disaster does occur; they don't know that appropriate coverage is available (or, in the case of a floodplain property securing a federally backed mortgage, that it is required); or they do not know about the hazard or cannot accurately assess their exposure to it.

The availability of federal disaster assistance is often cited as a reason why floodplain residents do not purchase flood insurance. The prevailing public impression is that federal disaster assistance is generally equivalent to the financial protection provided by hazard insurance. This is simply not true. Except in the case of special initiatives, such as the buyout program after the 1993 Midwest flooding, the primary federal assistance provided to property owners after a disaster is low-interest loans from the Small Business Administration (SBA). Disaster victims who are deemed unable to repay an SBA loan can receive an Individual and Family Grant from the Federal Emergency Management Agency (FEMA). However, each grant is limited to $12,900, and most are much smaller (the average awarded by FEMA in recent disasters is $3,000). Furthermore, they are intended only to meet reasonable needs and necessary expenses, not to make the victim "whole" again. However, despite the fact that by any standard insurance protection is preferable to either a loan or a grant, the public misperception persists and has proved difficult to correct.

Government policy can have significant implications for mandates on the purchase of disaster insurance. One the one hand, such mandates can help combat adverse selection and expand the pool of properties that are insured. This contributes to risk diversification and the capacity of the system to accommodate large catastrophic events. On the other hand, coverage mandates can significantly increase political pressure on insurers and regulators to make coverage available at suppressed prices, and cause regulatory interference with market forces. It may be preferable to encourage the purchase of insurance, short of mandates, and use voluntary mechanisms to ensure the maximum availability and purchase of coverage at the lowest possible price. Of course, this could require the government to be less generous in providing postdisaster aid, which would be a difficult political task indeed.

The Supply of Disaster Insurance

The insurance industry is encountering serious problems in providing insurance for properties located in areas subject to catastrophic losses, particularly those exposed to hurricanes and earthquakes. The problems fundamentally arise from the fact that many insurers now realize they do not have the resources to pay for a so-called worst-case event in those high-risk areas. This is because low-frequency but costly events do not provide a statistical base adequate for sufficiently accurate projections. This is further complicated by the difficulty of aggregating adequate capital over a number of low-loss years in highly competitive financial markets. In addition, insurers face challenges in setting rates based on risk because of the pressures inherent in the current insurance regulatory system.

The realization that resources may not be available to cover claims in a catastrophic event has come only recently in the case of hurricanes, which have been routinely covered by residential and commercial insurance policies for many years. Before 1988 the insurance industry had never experienced a loss greater than $1 billion from a single event. Since that time there have been 15 disasters exceeding $1 billion in insured losses. The pivotal wake-up call was provided by Hurricane Andrew, which generated $15.5 billion in insured losses even though it bypassed the most heavily developed parts of the Miami metropolitan area. Natural disasters costing $50 billion to $100 billion in insured losses are now possible and even likely. Losses of these magnitudes could create unmanageable problems for property owners, mortgage lenders, the insurance industry, and the affected communities.

Insurers confronted by catastrophic loss situations have tried to deal with them by diversifying their book of business to avoid overconcentration in a given state or region, by purchasing reinsurance to spread the risk more broadly, and by charging higher premiums in high-risk areas to cover catastrophic losses. In Florida and California, two of the highest-risk areas, emergency regulations and other laws have hampered insurers' pursuit of those solutions. Some companies have concluded that the resulting risk of insolvency is unacceptable and have attempted to withdraw entirely from those states. Others have stopped writing any new business there until their excessive risk exposures can be reduced.

Insurers also are concerned about questions of equity when faced with the likelihood that catastrophic losses arising from particular states or areas within states will drain off dollars collected to pay losses in other regions and other lines of insurance. For example, the Northridge earthquake produced insured losses of more than $12.5 billion. But only $1 billion in premiums were collected specifically for earthquake shake damage in California in 1994. The Northridge earthquake clearly was subsidized by premiums collected for other purposes and by insurance policies written in other states.

State insurance laws require that premiums not be excessive, inadequate, or unfairly discriminatory. Unlike government welfare plans, private insurance does not deliberately transfer wealth from one state to another or from one class of policyholders to another. This may happen in the short run, particularly when large catastrophes occur, but over time each group and geographical area is supposed to pay its own way.

Many of the political arrangements created recently at the state level run counter to this principle. The most common mechanism is a state-mandated pool, which serves as a market of last resort for property owners when coverage is not readily available from private insurers. Since the pools typically do not charge a premium high enough to cover the catastrophic loss potential of the properties involved, they subsidize the people living in high-hazard areas and impose the excess costs on people residing elsewhere. Moreover, these state pools do not eliminate the problem of catastrophic losses. Private insurers in those states remain liable, on a market share basis, for the net losses generated by the state pools. Thus, any increase in voluntary business carries with it an increase in the insurer's share of the adverse results of the pool. This creates a disincentive for existing insurers to remain in those states or for new companies to establish operations there.

Insurance Regulation

Regulation influences the supply of disaster insurance by controlling insurers' entry to and exit from insurance markets, capitalization, investments, diversification of risk, prices, products, underwriting selection, and trade practices. In theory the job of state regulators is to protect the public from fraud and imprudent practices that threaten insurance companies' solvency and to ensure fair market practices. However, public policy is not forged in a political vacuum, and regulation increasingly has been influenced by voters' perceptions and preferences on how the cost of risk should be shared among different groups. In the process, insurers have largely lost both the freedom to choose the exposures they are willing to insure and the freedom to charge premiums based strictly on a structure's loss potential.

Regulators, too, are faced with a difficult challenge—that of assuring an adequate supply of "affordable" insurance coverage at a time when many insurers are seeking to decrease their disaster exposure and increase their prices for the catastrophe component of that risk. Resolution of this dilemma could have substantial implications for the economies of many disaster-prone areas and their residents.

Mitigation and Insurance

Insurance itself is not considered a mitigation measure because it redistributes rather than reduces losses, but a carefully designed insurance program can encourage the adoption of loss reduction measures by putting a price tag on the risk and creating financial incentives through rate discounts, lower deductibles, and higher coverage limits. There are no easy explanations and no easy solutions to the problem of mitigating and insuring against natural hazards. There is, however, an increasing recognition by those in the insurance sector, the model code organizations, and government that a program must be developed that will address these issues. There are four principal means by which the insurance industry can facilitate mitigation.

1. *Education and information.* A major role for the insurance industry is to engage in educational programs designed to enlighten individual property owners about the risks they face and the mitigation actions they can take to reduce their chances of loss. An informed property owner is more likely to engage in risk reduction and to purchase insurance (see Mileti and Fitzpatrick, 1993).

2. *Participation in the model code process.* The insurance industry must become an active participant in the code development process (as it did before the 1980s). Through this process the industry can make its case for better codes to reduce property losses from natural hazards. The insurance industry has as much at stake in the outcome of these processes as do the homebuilder associations, real estate interests, materials suppliers, and local code officials. After model codes are improved, the insurance industry can actively encourage communities to adopt and enforce them.

One of the insurance industry's most significant concerns is that building codes historically have been designed for life safety and contain few if any provisions for reducing damage to property. This view persists today among model code groups and local building officials, the groups most involved in writing the codes and enforcing them.

3. *Offering financial incentives.* The most frequently suggested financial incentives are insurance premium reductions, changes in the amount of the deductible, and changes in coinsurance schedules, which reflect the changes in risk resulting from the implementation of a mitigation program. In the case of premium reductions, individuals would compare the reduction in premium offered with the estimated cost of mitigation action and decide whether the mitigation measure is beneficial based on a perception of the risk. It should be noted that, before a premium can be reduced (as an incentive), it must develop sufficient funds to pay for losses. In addition, any premium incentive must be approved by state regulators (except for the NFIP, which as a federal program is not subject to state regulation). Deductibles and coinsurance involve risk sharing and are designed to encourage property owners to protect against small losses, which also benefits insurance companies by reducing the expense of dealing with small claims.

The high front-end cost of mitigation versus a premium reduction spread over many years may weaken the financial incentive. It could, however, be bolstered with noninsurance incentives that would yield benefits to the property owner who holds a policy in the shorter term (e.g., a waiver of property taxes that would be derived from the increase in the property's value as a result of the retrofit, a waiver of sales tax for materials used in the retrofit, or a waiver for building permit fees.) Innovative financing programs could help, like long-term loans tied to mortgages or awarding a mitigation seal of approval to raise the price of a house at resale.

4. *Limiting the availability of insurance.* Property owners would be most likely to implement mitigation measures if insurance were not available until after the property had been built or retrofitted to an acceptable standard. Given sufficient market penetration, the application of market forces that make the availability of financing and insurance for buildings dependent on their meeting certain high mitigation standards should help motivate builders to build to a higher standard and owners to retrofit existing properties.

Summary

Dealing with natural disasters of the magnitude now predicted, and to nest insurance within sustainable hazards mitigation, will require new policies to encourage or require property owners to take cost-effective mitigation efforts and to provide compensation to cover the losses. Private insurance can pursue both of these sustainability strategies, but the problems may be too large for the insurance industry alone to handle. Public programs such as disaster relief also have a role to play, but their cost is becoming increasingly burdensome to taxpayers and they offer no incentive to undertake mitigation. There is an opportunity to utilize insurance as an important part of a hazards management program that would encourage and enforce cost-effective loss reduction measures. The limited use of such measures on existing structures in the United States indicates that new approaches must be developed by key stakeholders such as the insurance industry, financial institutions, state regulators and insurance commissioners, the building industry, inspectors, and real estate developers for reducing losses from these catastrophic events.

The new strategy should include improved estimates of risk, certifications of damage resistance, heightened enforcement, a policy decision about whether and how to subsidize mitigation and/or disaster losses for low-income households or others, and additional ways to protect the insurance industry against insolvency. Progress will require the direct involvement of government at all levels to link insurance and mitigation. Technical assistance, especially in risk assessment and testing and evaluating new mitigation methods, will provide the insurance industry with improved understanding of the risks.

PREDICTION, FORECAST, AND WARNING

The United States has no comprehensive national warning strategy that covers all hazards in all places. Instead, public warning practices are decentralized across different governments and the private sector. Uneven preparedness to issue warnings exists across local communities; hence, people are unevenly protected from the surprise onset of natural disasters. Without changes in this situation, inequities will grow larger, and the gains made in saving lives over the past decades may well be reversed.

Warning systems detect impending disaster, give that information to people at risk, and enable those in danger to make decisions and take action. This definition is simple, but warning systems are complex, since they link many specialties and organizations—science (government and private), engineering, technology, government, news media, and the public. The most effective warning systems integrate the subsystems of detection of extreme events, management of hazard information, and public response and also maintain relationships between them through preparedness.

Hazard-Specific Knowledge

Since the first assessment (White and Haas, 1975) was completed, there have been significant improvements in forecasts and warnings for some hazards but only marginal improvements for others. Forecasts for flood, hurricanes, and volcanic eruptions have improved most significantly, and public dissemination of warnings has improved the most for hurricanes. A 100 percent reliable warning system does not exist for any hazard.

Flood

Flood forecast and prediction capabilities evolved slowly during the 1970s and 1980s; more recent advances could have a major impact on forecasting (e.g., systems under development at the National Weather Service [NWS] include the NEXRAD Doppler radar, the Advanced Weather Interactive Processing System, and the Automated Surface Observing System). Floods are forecast by hydrological models that estimate flood conditions based on predicted or measured parameters, through physical detection systems, or a combination of the two approaches. Flash floods remain difficult to predict. Public flood warning dissemination and integration capabilities have improved only marginally.

The NWS is the only federal entity with a mandate to issue flood warnings, but other groups also are involved—for example, local floodplain managers, the U.S. Army Corps of Engineers, the U.S. Bureau of Reclamation, and private forecasters. Locally based detection systems also exist in many communities.

Tornado

There have been significant improvements in warnings for tornadoes over the past two decades. In 1978 warnings were issued for 22 percent of tornadoes; the average lead time was three minutes. In 1995 the percentage had risen to 60 and the lead time to almost nine minutes. The nation is moving from "detected" warnings to an era of "predictive" tornado warnings. Tornado prediction has improved over the past 20

years because of geosynchronous satellites (GOES), improved use of radar, better training for forecasters, and improved local storm spotter networks and awareness campaigns. Models of how tornadoes form have matured from simple ones to those using parent circulation (mesocyclone) at midlevels in thunderstorms.

Dissemination of tornado warnings to the public and emergency managers is significantly different now compared to two decades ago. One change has been the growth of the private-sector weather information industry. Another advancement has been the growing ability of the news media to quickly get warnings to their audiences. Finally, meteorological support companies now provide many services that allow meteorologists to create their own forecasts and on-air displays.

Hurricane

The National Hurricane Center (NHC) in Miami, part of the NWS, is responsible for predicting hurricane behavior and issuing warnings. Other entities also make hurricane forecasts, but only NHC forecasts are

A tornado tore a hole in downtown Clarkesville, Tennessee on Friday, January 22, 1999, destroying historic buildings and knocking out power in much of the city, but causing only minor injuries. It ripped the roof off the courthouse and knocked down most of a church. Photograph courtesy of AP/ Wide World Photos.

disseminated through the NWS's centralized computer system. Private meteorological firms provide operational hurricane forecasts to private industry clients and state and local governments.

Predicting hurricane behavior has three components: (1) collecting accurate data about the hurricane itself—location, wind and pressure profiles, speed and direction—and about the surrounding atmosphere; (2) anticipating changes in the associated meteorological environment; and (3) understanding why hurricanes behave the way the do. Hurricane data are provided by satellites, reconnaissance aircraft, buoys, ships, and coastal radar and are crucial for identifying the hazard, detecting trends, and establishing initial conditions required by predictive models. Visible and infrared satellites provide useful indications of a hurricane's center location, central pressure, and wind velocity. However, they are not accurate enough for some predictive models. Hurricane forecasts are most accurate in the Gulf of Mexico because there is a denser network of data about the atmosphere there (Sheets, 1990).

Hurricanes are part of large air masses. One aspect of forecast difficulty is anticipating the changes in those masses and determining how

TABLE 6.1 Maximum Hurricane Strike Probabilities for Forecast Time Frames

72 hours	10%
48 hours	13-18%
36 hours	20-25%
24 hours	35-50%
12 hours	60-80%

SOURCE: National Hurricane Center and Emergency Management Institute, 1995.

they influence a hurricane. Much effort is expended on discovering those influences and incorporating them into predictive models. Models often disagree, so forecasters consider the output from all models and then make forecasts. Official forecasts are three hours "old" when released.

The NHC generates a variety of forecast-related products. Public advisories are intended primarily for use by the news media. They give current hurricane conditions and indicate the general direction of a hurricane, if it is expected to strengthen, the peak storm surge expected, and the amount of predicted rainfall. The possibility of tornadoes may be noted, along with appropriate actions to be taken by the public. Public advisories may also include tropical storm watches and warnings.

A hurricane watch is issued for 300 or more miles of coastline when NHC forecasters believe a hurricane can strike land within 36 hours. When forecasters believe a hurricane can strike land within 24 hours, the NHC issues a hurricane warning, also for a broad stretch of shoreline. A tropical cyclone forecast/advisory contains specific forecasts regarding where, how strong, and how large a hurricane will be in 12, 24, 36, 48, and 72 hours, in addition to comparable current information about the hurricane. A strike probability forecast indicates the probability that the center of a hurricane will pass within 65 miles of a list of locations in certain time frames. As a hurricane approaches shore, forecasts become better. Table 6.1 indicates the largest strike probability values any place will have at certain periods from landfall.

Position (or location) forecasts are most accurate for time frames closest to the current time. Over the past 25 years, 24-hour position forecasts have been improving at an average rate of 1.1 percent per year (McAdie, 1996). Over the same period, 48- and 72-hour forecasts improved 1.5 percent and 1.2 percent, respectively.

Historically 90 percent of the people who have died in hurricanes have drowned in storm surges. The NHC uses a computer program named Sea, Lake, and Overland Surges from Hurricanes (SLOSH) to predict the height and areal extent of storm surges. Inundation scenarios are created and put into a SLOSH atlas, which depicts the maximum heights throughout a community for each storm category.

Evacuation planners anticipate the coastal area that will need to be evacuated because of the storm surge, how long the evacuation will take, and what accommodations must be provided to evacuees. Studies for such plans are funded by FEMA, the U.S. Army Corps of Engineers, NHC, and sometimes state emergency management agencies.

Drought

Droughts differ from other natural hazards in four ways: (1) there is no universally accepted definition of drought; (2) drought onset and recovery are usually slow; (3) droughts can cover a much larger area and last many times longer than most natural hazards; and (4) droughts are part of the natural variability of virtually all climatic regimes, rendering the entire United States at risk (Wilhite, 1993). These differences have prevented many state and local governments from establishing drought mitigation or contingency plans, including early warning and detection systems.

A reliable long-term forecasting model for droughts does not now exist. Scientists are striving to predict droughts by concentrating on teleconnections between large-scale atmospheric/oceanic anomalies and drought. In order for a meteorological drought to occur, usual precipitation patterns must be disrupted. Forecasting meteorological conditions is only part of predicting a drought. The severity of drought also depends on the moisture content of the soil and the general health of vegetation.

Remotely sensed data can be used to determine both plant health and soil moisture content. The National Oceanic and Atmospheric Administration's (NOAA's) advanced very high resolution radiometer data have been used to produce an index of vegetation conditions, which has brought particularly good results in drought detection and can contribute to early warnings. For short-term forecasts the NWS already provides 3- to 5-day and 6- to 10-day precipitation forecasts. In addition, 30- and 90-day outlooks are issued.

Improvements in drought prediction and forecasting are developing on three fronts. First, scientists are gaining a better understanding of the interactions and feedback mechanisms between physical systems and the causes of drought. Second, there have been continued technological advances in meteorological/climatological instrumentation. As the accuracy, dependability, and durability of these instruments improve, so will the quality of the data gathered. The combination of more accurate and consistent data and better models should result in more reliable predictions and forecasts. Third, there is a more widespread recognition of the importance of integrating physical and social parameters at the community level. Currently, little is known about how drought forecast information would be used by decisionmakers.

A national drought policy and plan was initiated in 1987. This plan was designed to help state governments prepare for droughts. Ten essential but flexible steps were delineated (Wilhite, 1993). An early warning system was one of the mitigation measures suggested, but it has yet to be established.

Snow Avalanche

The first regional snow avalanche forecast center in North America was founded in 1962. Today there are nine such centers—two in Canada and seven in the United States. They are responsible for monitoring and forecasting avalanche danger in backcountry or highway corridor areas ranging in size from 1,000 to 100,000 square kilometers.

A recent survey of the centers (Williams, 1996) determined that they use the latest technologies to gather and receive data and to disseminate forecasts—automated remote data stations, manned observation sites, stability tests, and reception and dissemination via modem, fax, mail, and the Internet. However, few technological aids are used in data analysis and decisionmaking. Rather, all of the centers rely on conventional avalanche forecasting—using measurements, observations, experience, intuition, and knowledge of prevailing local terrain, weather, and snowpack conditions.

The state of the art in snow avalanche forecasting is dependent on the amount and type of research being conducted. In the United States very little research is being done because of a lack of federal and state funding. Consequently, U.S. forecast centers rely on research findings and technology imported from other nations. Among the technological advances, products, systems, and methods available for snow avalanche forecasting are remote automated weather data systems; improved NWS numerical forecast models; nearest-neighbor models; expert systems; avalanche hazard indexes for highways; stability tests such as shovel shear, rutschblock, compression, and stuffblock; geographic information systems (GISs); e-mail; and the Internet.

Technological advances have greatly improved the methods by which centers gather and receive data and disseminate their forecasts, but new technology is scarcely used for analysis and decisionmaking. Centers rely almost totally on conventional methods of avalanche forecasting based on experience, intuition, and local knowledge for many reasons: conventional forecasting is a proven method; regional forecast centers must analyze large amounts of weather, snowpack, and avalanche data gathered over large tracts of mountainous terrain, which compounds the problem of trying to use numerical techniques to aid analysis; the decisionmaking process in avalanche forecasting does not lend itself to modeling; and there is no budget for research and development of new technology to produce a useable and sophisticated computer aid for conventional forecasting.

Wildfire

Forecasting fire behavior depends on predicting the interaction of topography, fuels, and weather, especially to predict quickly how changing weather will affect a fire. Warning systems for the prediction of wildfire behavior include fire danger rating systems and fire behavior modeling systems. Rating systems usually attempt to predict the probability of ignitions. The system will then try to predict the potential fire behavior for a relatively large area. Fire management organizations use this information to plan readiness levels of fire-fighting staff and restrictions in human uses of wildlands and equipment.

At present all agencies in the United States use the National Fire Danger Rating System (NFDRS). The system is also used by local fire agencies that oversee the urban/wildland interface. Recent improvements in the NFDRS include the use of remote automatic weather stations. These stations can be located anywhere and will transmit the data necessary to calculate fire danger as well as give local weather for other planning such as prescribed burning. Every three hours a station is triggered by the GOES satellite to transmit data to it. The satellite then transmits the data for calculation.

Current shortcomings in the system stem from three sources. The first is the limited ability to understand the basics of fire dynamics in complex fuel, topography, and weather systems. A second source is the broad nature of fuel models and the large areas over which predictions must be made. By far the greatest source of failure, however, is the inability to predict weather accurately enough and far enough in advance over the complex terrain that is often involved.

Fire behavior prediction in the United States is presently based on the BEHAVE system (Andrews, 1988), which draws on the fire behavior models originally published in 1972 and improved upon over the past 20 years. The program will provide the rate of spread, fireline intensity, heat release per unit area, and flame length. The operator can also ask for the distance that the fire will spread in given time periods. Recent work has led to the real-time prediction of a moving fire and given personnel better information on how and where to fight a fire and make better evacuation decisions.

Earthquakes

Programs directed at predicting earthquakes have had mixed success. Through statistical analysis of earthquakes worldwide, the frequency of different-magnitude quakes across the globe can be estimated. The monitoring of global seismicity also makes it clear that certain areas are much more prone to quakes than others. For example, 90 percent of the world's earthquakes occur on the boundaries of large tectonic plates. Along a single plate boundary, however, there can be considerable variability in the size and frequency of significant earthquakes. Parts of the San Andreas fault accommodate the relative motion of the North American and Pacific plates without earthquakes through aseismic slip; other sections of the fault have experienced several large or major quakes during recorded history. In general, intraplate earthquake sources and processes are even less well known. Thus, a better understanding of the relationships among plate tectonics, regional stresses, and earthquake sources is needed.

Scientists are making progress in understanding earthquake genesis and growth. Recent observations suggest that conditions favoring the growth of large, potentially destructive earthquakes are fundamentally different from those that lead to smaller, more common events. If so, careful geological and geophysical monitoring might someday detect the conditions that signal imminent earthquake risk (see Ellsworth and Beroza, 1995).

Local geology and topography may also have a role in whether larger, less frequent quakes—or smaller, more frequent ones—are to be expected on a fault. Advanced models of rupture propagation, additional geophysical data, and additional seismological data from newer broadband high-dynamic-range instruments will likely aid in understanding how surficial and subsurface fault characteristics affect rupture and maximum magnitude.

The standard approach to developing a prediction capacity hinges on the earth's providing recognizable signals of impending quakes. Theoretical and laboratory studies indicate that there should be preliminary phases before rupture. Potential earthquake precursors include foreshocks, changes in the groundwater table, other hydrological or hydrothermal phenomena, deformation of the earth's surface, changes in the rock's electrical conductivity or magnetic properties, and changes in seismic wave properties through the area in question. In the past, such phenomena have been observed in the field but not consistently.

Some advances in the detection of earthquake and seismological data, as well as dissemination, have been made. The CUBE (California Institute of Technology-U.S. Geological Survey [USGS] Broadcast of Earthquakes), in southern California, and REDI (Rapid Earthquake Data Integration) in northern California, are programs that provide rapid information, including the magnitude, location, depth, and other data on California earthquakes. Data from the University of California at Berkeley, Caltech, and the USGS seismic networks are automatically processed for earthquake parameters, and the information is distributed by pager, e-mail, and the World Wide Web. Recipients of these earthquake data include public and private agencies in emergency services, utilities, and lifelines.

TriNet is a five-year $20 million project to develop a state-of-the-art digital seismic network in southern California. The TriNet partners include the USGS, the California Division of Mines and Geology, and the California Institute of Technology. The project will result in tangible seismic safety benefits that include accurate and reliable locations and magnitudes of earthquakes and maps of the regional distribution of ground shaking within five minutes of a significant earthquake. TriNet will also develop a pilot earthquake early warning system for southern California.

The first U.S. effort directed at earthquake prediction was located near the central California town of Parkfield, adjacent to the San Andreas fault. The Parkfield prediction experiment was begun in 1985 after analysis of previous earthquake occurrences on a particular fault section indicated that a repeat event would occur near the end of the decade (Bakun and Lindh, 1985).

In November 1988, two USGS scientists submitted to the National

The climactic eruption of Mount St. Helens, May 18, 1980, at about noon. The maximum height of the ash and gas column was about 12 miles. Photograph by Robert M. Krimmel.

Earthquake Prediction Evaluation Council (NEPEC) data indicating that the chances of an earthquake were very high. Within three months both NEPEC and the California Earthquake Prediction Evaluation Council had endorsed the scientists' prediction. Thus, the Parkfield experiment became the first scientifically credible long-term earthquake prediction. The director of the USGS issued a formal public forecast of the quake in April 1985, stating that there was a 90 percent probability of a magnitude 5.5 to magnitude 6.0 earthquake sometime between 1985 and 1993 in the Parkfield area. It also stated that a 10 percent probability existed for a magnitude 7.0 quake. The release of this forecast became a national media event and precipitated a media campaign in central California involving newspapers, radio, and television that lasted years. In 1988 the California Governor's Office of Emergency Services published a detailed brochure and mailed it to 122,000 households at risk. It covered information about the earthquake hazard, the prediction, a possible short-term warning, and how to take action. But the expected characteristic earthquake never happened. Further analysis showed that, while the successive repeat of similar but not identical quakes might be expected on individual fault sections, the amount of time between them may be highly variable. Confidence in predictors based on estimates of recurrence intervals has decreased; scientists are more sanguine about the possibility of identifying one or more of the "red flags" described above.

Volcanoes

Volcano hazard forecasting and prediction includes forecasting explosive events and assessing volcanic hazard. In the past 20 years considerable progress has been made in identifying the extent and magnitude of hazards at high-risk volcanoes. But scientists have had mixed success with predicting the timing and magnitude of volcanic eruptions. Predictions have been most successful for volcanoes with frequent eruption cycles and most difficult for large caldera systems with low recurrence intervals. An obstacle to prediction advancement is lack of instrumentation and monitoring on most active volcanos around the world. Although efforts have been made to integrate hazard assessment into emergency planning, there is not much evidence of integrated volcano warning systems in potentially hazardous areas in the United States. A current trend is toward developing planning scenarios based on the information in a volcanic hazard assessment, which could be used to design a warning system.

The monitoring activities associated with prediction fall into the categories of seismic monitoring, seismic tomography, ground deformation monitoring, electromagnetic monitoring, and geochemical monitoring. The greatest progress has been made in recent years in understanding volcanic structure and eruption dynamics. Seismic tomography techniques have led to better understanding of the subsurface structure of volcanoes, and modeling of stress-strain buildup as an eruption precursor has advanced considerably. Some success has been achieved with gas emanations modeling. In contrast, geoelectric monitoring and geomagnetic monitoring have produced inconsistent results.

Tsunamis

Tsunamis are instantaneously generated traveling waves of water. Most tsunamis occur in the Pacific and Indian oceans, but they also occur in almost all large bodies of water. All great oceanwide tsunamis are generated by large subduction-zone earthquakes, but large local waves can be generated by submarine and subaerial landslides, various volcanic processes, and other phenomena. The characteristics of their generation and propagation provide the means for detecting tsunamis and issuing warnings. Because the causative earthquakes are very large and near the coast, they can be recorded around the earth in minutes. It is possible to detect them, locate the epicenter, and calculate the magnitude in about 30 minutes, well before the arrival of tsunami waves from remote sources (it takes from several hours to a day for a wave to reach the opposite coast) but not necessarily before the arrival of a locally generated tsunami.

Improvements in the past 20 years in warning and forecasting technologies for tsunamis have been mainly in communications and procedures. This includes the real-time satellite communication of analog seismic and tide gauge data from stations around the Pacific to warning centers. Use by the Japanese and the National Earthquake Information Center of computers that can automatically scale incoming data and compute locations and magnitudes have speeded up the warnings. The recent development of deep-ocean gauges promises to provide rapid data on wave heights from the open ocean, uncontaminated by shoaling phenomena as are gauges in harbor locations. So far these gauges are operated off the Shumagin Gap area in Alaska, the state of Washington coast, and Hawaii. Their more general deployment will result in substantial improvement in the system.

Computer models have been of limited use in predicting runup for inundation mapping, partly because of the lack of good initial data. Models should be useful in reconstructing the source conditions, but again, their success has been limited. Predicting the height of waves from remote sources at specific localities also should be possible with models, but so far it has not been done.

Landslide

Landslide prediction and forecast in the United States has traditionally focused on identification of landslide-prone areas and potential slide locations. Landslide hazard studies are carried out in areas with repeated high economic losses from landslides. Typical studies include regional identification of landslide potential, evaluating the kinds and intensities of landsliding, determining their areal distribution and frequency, and studying landslide processes. To date, such studies have produced deterministic maps of landslide potential. These provide useful information for educating people about the hazard but are not particularly useful for issuing forecasts. Probabilistic landslide maps have not been extensively produced but likely will be in the future. They require the development of an historic frequency of landslide occurrence as well as a better understanding of rainfall intensity duration thresholds.

Progress has been made in predicting and forecasting both conditions that may lead to an alert for increased landslide potential in a region and the triggering of landslides at a specific site. Predicting and forecasting landslides depends on comparing precipitation and snowmelt forecasts and real-time observations to threshold values associated with the triggering of a landslide. Thresholds for rainfall-induced debris flows have been successfully developed for the San Francisco Bay area, and thresholds for snowmelt-induced landsliding and debris flows have been developed for the central Rocky Mountains.

Landslide prediction and forecast systems are also being used in conjunction with both seismic- and volcanic-induced landsliding. Work in southern California focuses on the near-real-time prediction of earthquake-induced landslides by combining measurements of strong ground shaking during an earthquake with GIS-based datasets on topography, geology, engineering strength, seismic intensities, and historic landslides. Work on volcano-related landslides characterizes the strength of volcanic rock, models the effects of destabilizing thermal pressurization, and models volcanic edifice stability.

The Landslide Hazard Program (LHP) of the USGS is a Congressionally authorized program dedicated to the reduction of landslide damage. The USGS also has been delegated the responsibility of providing landslide warnings. Until recently there has not been a mechanism to promote the dissemination of landslide prediction and forecast information to people in hazardous areas. The USGS National Landslide Information Center, in cooperation with the LHP, is developing a communications center for issuing advisories, press statements, and other information about landslides.

Technological Hazards

Technological hazards are products of industrialization that pose a health or safety threat to humans. Typically this includes nuclear materials, chemicals, and hazardous materials, including explosives, oil and gas products, and wastes. They can occur at stationary facilities or during transportation. Warning systems are much more feasible for fixed facilities, but their feasibility varies greatly according to the industry and location in the country. The prediction and forecast of technological hazards involves a complex process that is specific to the technology and/or materials involved. The basic approach is to estimate an accident sequence consisting of an initiating event, a release mode and a quantity and to model the dispersion of the release as it occurs—either visually or with instrumentation.

For this second assessment of hazards, technological hazards were viewed as secondary effects of natural events. Two approaches have been used to further understanding of the conditions under which natural events will lead to releases of hazardous materials. First, case studies and reviews of historical events have been conducted. These have led to fairly detailed inventories of the type of materials released in various events, although it has been argued that improved methods and much more careful study are needed. The second approach has been to model such releases using probabilistic risk assessment methods that typically include accident scenarios involving earthquakes, floods, tornadoes, lightning, storm surge, and high winds.

Emergency response requirements for chemicals and hazardous materials fall under a number of regulatory programs, particularly the Resource Conservation and Recovery Act, the Superfund Amendments and Reauthorization Act Title III, and the Clean Air Act's Risk Management Planning. All require some level of hazard disclosure by the material owner to communities to assist with a local warning system. These programs and their associated planning guides are vague about specific needs for alert and notification requirements. Another federal program is the Chemical Stockpile Emergency Preparedness Program (CSEPP). The CSEPP planning guidance contains detailed design criteria for a state-of-the-art alert/notification system. Warning systems developed at the eight CSEPP sites meet or exceed those for nuclear power plants.

Most communities with chemical or hazardous materials hazards do not have special warning systems. Most of the existing systems were developed primarily by individual private companies and communities as cooperative efforts. Research has shown, however, that in most chemical or hazardous materials accidents the prime responsibility for issuing a warning falls on local emergency response organizations, which are the first to arrive at the scene of a spill. The primary warning problems these organizations face are identifying the hazardous materials involved in an incident, determining the threat presented, and then deciding who to warn and what to tell them. Some communities have plans to guide this activity, but most incidents require ad hoc responses.

The Three Mile Island nuclear accident and the Bhopal, India, chemical accident revealed the need for monitoring and detecting nuclear power plants and many large chemical plants. But not all chemical and hazardous materials handlers have developed the ability to monitor conditions or predict or detect releases. There have been significant improvements over the past 20 years in problem recognition and identification, identifying options for protective action, and improving decisionmaking by development of decision support systems. A variety of dispersion models, decision support tools, integrated models, and emergency management information systems have evolved in various chemical response programs. As such systems develop, fuller integration of technological and natural hazard decision support efforts can be anticipated.

Evacuation has always been the preferred means of protecting the public from an accident. In the past 20 years much work has been done on alternatives to evacuation. Shelter in place—or protecting people where they are, without evacuating them—is accomplished by shielding the public from exposure pathways—vapors, aerosols, and liquid contamination. Shelters may be congregate (for many people) or individualized (a home). Shelters may be existing structures, with or without upgraded protective measures, or facilities specifically designed to protect from toxic chemicals.

Cross-Cutting Knowledge

A great deal of forecast and warning-related knowledge applies to several or all hazards. Much, but not all, of this knowledge resides in the social sciences.

Nation-Wide Approaches

The Civil Defense Warning System (CDWS) and the Emergency Broadcast System (EBS) were developed to warn of enemy attack, accidental missile launch, or radioactive fallout. The CDWS combined national, state, and local resources, but the heart of the system was the National Warning System (NAWAS). Operated by FEMA, it consisted of a series of nationwide, dedicated, 24-hour telephone lines, two national and 10 regional warning centers, primary warning points, state warning points, extension warning points, and duplicate warning points. The NAWAS was supplemented by state and local civil defense warning systems that transmitted warnings to officials and the public. State civil defense offices are usually linked to other state agencies, county sheriffs, and civil defense agencies. Local civil defense officials transmitted warning information to institutions and to the general public, primarily through local television and radio stations. EBS could be activated at the local level for community emergencies. The National Oceanic and Atmospheric Administration has developed Weather Radio to provide warnings of severe weather through commercially available tone-alert radios. Broadcast stations exist around the country, each serving a 40- to 60-mile radius.

In 1994 the Federal Communications Commission announced the creation of the Emergency Alert System (EAS) to replace the EBS. EAS will cover both national and local emergencies and is designed to take advantage of current digital communications technology. All commercial broadcast stations and cable companies are required to participate in the system. Some of the features of the new system will include multiple alerting sources, remote operations, and targeting of specific geographical areas. The EAS and NOAA Weather Radio program are being integrated so that the EAS can activate the tone-alert radios. Except at the local level, EAS is not tied to any other hazard warning systems.

Warning Response

A great deal is known about how and why people (individuals, families, and organizations) respond to warnings. This knowledge has already been summarized (see Drabek, 1986; Lindell and Perry, 1992; Mileti and Sorensen, 1990) and key points are presented below.

Coordination and communication between the different organizations that are part of a warning system are essential to the issuance of timely public warnings. The conditions that facilitate and/or undermine coordination and communication between warning system organizations are well defined. Simply stated, coordination is maximized when organizations know what they and other organizations are supposed to do, know who in the organization is to do it, have designated and understood communication ties to other organizations in the system, and maintain flexibility. Communication problems, owing to both equipment and human failure, are the most significant causes of poor warning dissemination.

A fairly thorough understanding of warning compliance has been developed by social science researchers. The focus of their research has been on whether or not people evacuate when advised to do so. In contrast, little work has been conducted on how people choose protective actions. Nor have individual variations in response to warnings been explained, such as why some people act immediately and others delay (Sorensen, 1991).

Warning response is a process with several stages: (1) hearing the warning, (2) believing the warning is credible, (3) confirming that the threat does exist, (4) personalizing the warning to oneself and confirming that others are heeding it, (5) determining whether protective action is needed, (6) determining whether protection is feasible, and (7) determining what action to take and then taking it (see Lindell and Perry, 1992; Mileti and Sorensen, 1990).

Both general and specific factors that affect public warning response have been identified. These include characteristics of the message sender, the receiver, the message itself, and the social context a person is in when a warning is received (Mileti and Sorensen, 1990). These factors are summarized in Table 6.2.

The chief way that warning response can be affected by emergency planning is in the design of the warning system, including the channel of communication, preevent education, and how the emergency message is worded. Incentives also can be offered to increase response, including information hotlines, transportation assistance, mass care facilities, and security to protect property left behind.

Much progress has been made on measuring and modeling warning dissemination and response (see Sorensen and Mileti, 1989; Lindell and Perry, 1992). The knowledge generated includes data on the time that decisionmakers take to reach a decision to issue a warning, the time it takes to disseminate a warning via different technologies and strategies, the time it takes people to reach a decision to act on a warning, and the

TABLE 6.2 Major Influences on Response to Warnings

Factor	Direction of Impact on Public Response	Empirical Support
Physical cues	Increases	High
Social cues	Increases	High
Perceived risk	Increases	Moderate
Knowledge of hazard	Increases	High
Experience	Mixed	High
Education	Increases	High
Family plan	Increases	Low
Fatalistic beliefs	Decreases	Low
Resource level	Increases	Moderate
Family united	Increases	High
Family size	Increases	Moderate
Kin relations (number)	Increases	High
Community involvement	Increases	High
Ethnic group member	Decreases	High
Age	Mixed	High
Socioeconomic status	Increases	High
Gender (female)	Increases	Moderate
Having children	Increases	Moderate
Channel: electronic	Mixed	Low
Channel: media	Mixed	Low
Channel: siren	Decreases	Low
Personal contact	Increases	High
Proximity to threat	Increases	Low
Message specificity	Increases	High
Number of channels	Increases	Low
Frequency	Increases	High
Message consistency	Increases	High
Message certainty	Increases	High
Officialness of source	Increases	High
Fear of looting	Decreases	Moderate
Time to impact	Decreases	Moderate
Source familiarity	Increases	High

SOURCE: Mileti and Sorensen, 1990.

time it takes to carry out alternative protective actions such as evacuating or taking shelter. Additionally, many other lessons have been learned from warning research since the first assessment (White and Haas, 1975). Some of the more significant lessons include that officials are often slow to reach a decision about issuing a warning, and slow decisions often prevent an effective public warning; most populations can be notified in about three hours or less without specialized warning systems; warnings are more slowly disseminated at night; new warning technologies, such as telephone ring-down systems, can achieve very rapid warning times; informal notification plays an important role in the warning dissemination process in most emergencies; the time people spend responding to a warning corresponds to an s-shaped (logistic) curve; and, surprisingly, the time required to evacuate a population is unrelated to its size.

Alert and Notification Technology

A public warning is often considered to have an alert component and a notification component. The alerting phase gets people's attention with sound or other sensory stimuli. The notification phase communicates the information. Significant improvements have been made in alert and notification technologies over the past 20 years. Mechanical sirens, which only provided an alert signal, have been augmented with electronic sirens that can also broadcast voices. Electronic signs, some of which can be remotely activated, are now available for use on highways. Special tone-alert radios that can be remotely activated (like NOAA's Weather Radio) can be installed in homes or businesses to provide indoor alerts. Cable overrides have improved the use of television as a warning mechanism. Telecommunication devices for the deaf and strobe lights have improved the ability to communicate with people with certain disabilities. Other advances include telephone ring-down systems, which either use computers to automatically dial sequential banks of telephones or switching equipment that simultaneously dials a large number of phones.

As the communications revolution advances, new warning technologies likely will be developed and commercialized. The EAS SAME technology has increased the efficiency of television and radio for warnings. This technology will eventually enable remote activation of consumer products such as car radios to receive messages being disseminated over commercial stations. Improved tone-alert radios are being developed. The National Aeronautics and Space Administration is developing an alert and notification system that would use satellites to activate individual pagers. Most of these newer technologies have not been systematically investigated in a field setting, so it is not yet known whether they will be more effective alert and notification techniques.

Effectiveness

Much has been learned since the nation's first assessment of national hazards (White and Haas, 1975) about the timing of warning dissemination. The most significant debate on what constitutes a state-of-the-art alert and notification system came in an Atomic Safety Licensing Board (ASLB) proceeding on the Shearon Harris Nuclear Power Plant. In its final ruling the ASLB defined what constitutes "essentially 100 percent notification within 15 minutes in the first 5 miles of the Harris Emergency Planning Zone" (Atomic Safety and Licensing Board, 1986). The ASLB required the utility to establish that over 95 percent of the people within 5 miles of the facility would receive a warning within 15 minutes. In order to exceed the requirement, tone-alert radios were proposed for all households within the 5-mile radius.

Expedient warning remains a thorny problem for natural hazards emergency managers since some fast-moving events can provide no or only a few minutes of warning time. New warning technologies are needed for rapid warning dissemination.

Community Adoption

Little is understood about the adoption of community warning systems in the United States. There is strong anecdotal evidence that the NAWAS is poorly maintained at the local level and in most communities uses outdated control, alert, and testing technology. The new EAS strategy will only partially address problems of community adoption. As currently formulated, it is based on indoor technology and thus can only reach people already tuned in to the media.

Only one study has systematically investigated the adoption of warning systems; it was based on a national sample of communities and focused on chemical releases (Sorensen and Rogers, 1988). It showed that few communities used state-of-the-art communications equipment or warning system technology. The ability of the majority of systems to provide a timely alert and notification was highly questionable. Few communities had well-developed plans and procedures to guide emergency response. Notably lacking were organizational capabilities to make decisions.

Social Issues

The development and use of effective warning systems present several economic, ethical, and cultural issues.

Cost-benefit analysis. Two approaches have been used to estimate the costs and benefits of warning systems. First, average annual costs for warning preparedness are compared to the average number of lives saved by the system. In such analyses the benefits reaped can appear low for disasters that occur infrequently. A second type of analysis focuses on the potential for the infrequent catastrophe. This approach compares the cost of warning preparedness to the benefits of the system when the maximum credible disaster does occur. Most cost-benefit analyses of warning systems use both approaches. The results can vary widely across hazards as well as for the same hazard in different communities. Some decisions about warning system adoption and preparedness do rest on cost-benefit analyses. But often warning systems are developed based on humanitarian sentiments after a disaster, regardless of the outcome of a cost-benefit analysis.

Warning ethics. Warning systems are meant to serve the public good by saving lives and moveable property and by reducing injuries. Consequently, warning systems must influence and guide public behavior but not interfere with civil liberties. Debates over the ethics of warning systems have surfaced from time to time since the first assessment (White and Haas, 1975). For example, in the early 1970s a new alert device called DIDS was viewed as a warning breakthrough. The system activated radios to broadcast warning information. It was never adopted, however, because it was seen by many as a breach of privacy. Today, tone-alert radios are in place in many areas for some hazards, and the EAS technology is raising similar concerns about invasion of peoples' privacy.

Another frequently occurring issue of ethics has been whether warnings should simply advise the public about what protective actions to take or order them to take those actions. Contemporary consensus is that warnings should provide advice and recommendations. Sometimes this has meant standing by as some people decide not to evacuate and thus face almost certain death. For example, officials at Mount St. Helens knew that some residents refused to leave. Ethics questions continue to surface, and cannot be resolved readily.

Sources of information. Society is becoming exposed to greater amounts of information from an ever-larger number of sources, and people now receive warnings from more and more sources—some official, others not. Anecdotal cases of dual warnings in which an official source and another source are at odds have begun to surface. For example, in one case a local weather forecaster told people to ignore an NWS tornado warning—minutes later the tornado touched down; in another case a local television forecaster gave detailed storm track predictions on a street-by-street basis—far more detailed than could be supported by current scientific abilities. In yet another case a state agency provided hurricane storm track projections that differed from those issued by the NHC. Finally, a scientist made an earthquake forecast that was refuted by government seismologists.

These instances pose a basic dilemma for officials—how to get consistent and accurate information to the public given the fact that the government cannot regulate what unofficial sources say. The potential problems are clear: with inconsistent information people are less likely to believe a warning and more likely to waste time trying to get additional information to resolve inconsistencies, and many people will be less likely to take protective action in response to a warning.

Withholding warnings. The control and timing of public warnings continue to be thorny issues in emergency management. Several factors mentioned in the first assessment (White and Haas, 1975) still play a role in withholding warning information from an at-risk public. First, the unfounded but widespread belief that the public will become unnecessarily alarmed if warned about a low-probability but high-consequence event still prevails. This belief can sometimes result in reluctance to tell the public about an impending disaster until it is absolutely necessary, and even then some warnings are delayed, muddled, or suppressed. This reluctance to inform continues to affect both hazard detectors and emergency managers.

Second, sometimes warnings are still withheld because of concern over negative social and economic effects on the official or manager who will be responsible for coping with it and on society in general. In such instances there may be only a partial disclosure of information. This can seriously undermine warning effectiveness from the viewpoint of public protection. Additional information may well become public through unofficial sources, creating credibility problems for officials.

Multihazard Systems

If all hazards were the same, a generic warning system could be designed and used, but this obviously is not the case. There are six characteristics of hazards that affect one or more of the basic components of a warning system (detection, emergency management, and public response): (1) predictability relates to the ability to predict or forecast the impact of a hazard with respect to magnitude, location, and timing; (2) detectability refers to the ability to confirm the prediction that impacts are going to occur; (3) certainty is the level of confidence that predictions and detections will be accurate and not result in false alarms; (4) lead time is the amount of time between prediction/detection and the impact of the hazard; (5) duration of impact is the time between the beginning and ending of impacts in which warning information can be disseminated; and (6) visibility is the degree to which the hazard physically manifests itself so that it can be seen or otherwise sensed.

One kind of warning system will not work for all hazards or in all situations. But some events with similar characteristics may be able to use the same warning implementation strategy. In any case, hazard-specific knowledge must be incorporated into any warning system. A tiered scheme in which some components are shared across hazards but some are hazard specific may be the best approach. Any warning plan would address the warning system, organizational principles, and the basic public response process. The plan would then be specified or tiered into unique implementation procedures for each of the different hazard types that a community faces. Finally, unique hazard-specific information and site-specific conditions would be annexed to the plan.

Links to Sustainability

There is little doubt that improvements in prediction, forecast, and warnings have dramatically reduced deaths and injuries in the United States since the nation's first assessment (White and Haas, 1975). This is true for all hazards, but the same is unfortunately not true for many other parts of the world, particularly lesser-developed nations. Obviously, warnings can save lives and some moveable property and can reduce injuries. Beyond that, short-term (minutes to days) warning systems seem to have little direct bearing on sustainable development—or if links do exist, they have not yet been explored. Although they reduce deaths and injuries, warning systems have not been demonstrated to have any significant impact on reducing damage to social infrastructure or private property or on reducing economic disruption. In fact, short-term warning systems may hinder the movement toward sustainability by allowing long-term occupancy of marginal lands. For example, if people can return to occupy areas of high hazard because a warning system helped them avoid death or injury, the presence of a warning system may actually increase economic losses in the long run and jeopardize a sustainable economy. Again, the evidence is scanty and warrants further attention.

On the other hand, long-term warning systems (years to decades or longer) may have a major role to play in sustainable hazards mitigation. Long-term forecasts would provide local decisionmakers with some of the information needed to design their future communities. A certain amount of future losses would be part of any community's sustainable hazards mitigation plan because losses could never be reduced to zero. Long-term forecast systems would help redefine the risks that communities want to reduce, and information about the systems would be vital to the local planning process.

Conclusions

What has and has not been accomplished regarding warnings in the United States since the nation's first assessment was completed over two decades ago can be summarized in four phases:

1. *A national warning strategy*. The United States does not have a comprehensive national warning strategy. Warning practices are divided over different governmental entities and the private sector. For example, the new EAS being developed by the Federal Communications Commission is coordinated with the NWS but not with other public or private providers of prediction and forecast information, and different local communities vary greatly in the quality and likely effectiveness of in-place warning systems. The nation needs to develop a comprehensive model for warning the public, provide it to local communities along with technical assistance, and even out the degree of protection provided by warnings systems for all citizens.

2. *Improving warning systems*. Public alert systems can be improved with new hardware and technology, but diffusing existing technology and warning preparedness knowledge is a much bigger problem in the

nation today. Further technological advances will only increase the gap between practice and the state of the art. An exception would be the development of very inexpensive equipment that could be easily installed and maintained and that could rapidly alert and notify the public. The diffusion of SAME-enabled EAS warning devices into American households will likely be a slow process. Even when such devices are commercially available, few low-income residents will have them. Furthermore, the EAS cannot provide outdoor warnings.

Improvements to local warning systems are needed on two fronts. The first is the dissemination of information on low-cost or no-cost improvements. This includes improved procedures and management practices, which can result in a much better warning system without major financial expenditures. The second is the provision of funds for better communications and warning system equipment. Few communities have the funds to install new equipment and so will require technical assistance and/or cost sharing. Better local management and decision-making about the warning process are more critical than promoting more advanced technologies, although both would help. The most sophisticated equipment is relatively useless unless it can be used properly.

3. *Knowledge gaps*. The ability of a system to provide timely public warnings begins with monitoring the environment to detect hazards. Detection technology is readily available for some hazards but is only in a state of development for others. Technological capabilities also vary with respect to the amount of lead time provided and the "noise" in the detection signal. Monitoring technologies, which provide ongoing data about the physical system, are of equal importance. Again, monitoring coverage is fairly good for some hazards and poor for others, such as hazardous materials accidents. Complete coverage of the entire U.S. land mass, or even of all populated areas, has not been achieved for any hazard.

4. *Improved predictions, forecasts, and warnings*. Most advances in prediction and forecasting since the nation's first assessment in the 1970s have come from better monitoring, instrumentation, data collection, and data processing. Some of these have resulted from advances in theories and models, but no radical theoretical breakthroughs have occurred in the past 20 years. The ability to deliver warnings to the public—which means a thorough integration of the scientific component with an effective delivery mechanism—has a checkered record. Table 6.3 estimates

TABLE 6.3 Improvements in Prediction, Forecast, and Warning Integration

Hazard	Prediction/Forecast	Warning Integration
Flood	Some change	Not much change
Hurricane	Major changes	Major changes
Tornado	Some change	Not much change
Drought	Not much change	Not much change
Fire	Not much change	Not much change
Avalanche	Not much change	Not much change
Earthquake	Not much change	Major changes
Volcano	Some change	Not much change
Tsunami	Not much change	Not much change
Landslide	Some change	Not much change
Nuclear power	Major changes	Major changes
Hazardous materials/chemicals	Major changes	Not much change

the relative improvement in prediction/forecast and warning integration over the past 20 years. Even given the natural uncertainty in the behavior of hurricanes, improvements in prediction and forecasting capabilities and the ability to graphically present scientific information and warnings for that hazard have been exemplary. This is the case for nuclear power as well, although the impetus for that improvement came from regulatory requirements. Some advances have been made in predicting, detecting, and forecasting floods, tornadoes, volcanoes, landslides, and chemical accidents, but these improvements have yet to be integrated into warning dissemination. Earthquakes represent a unique case: while dramatic improvements have been made in integrating the warning process, our ability to predict earthquakes has not improved. Finally, four hazards have shown little change in either prediction/forecast or warning integration: droughts, wildfires, snow avalanches, and tsunamis. There is much room for improvement in the next 20 years.

ENGINEERING

Although efforts to reduce mortality rates from hazards and disasters over the past several decades have largely been successful, there is no way to determine how much of that reduction is attributable to state-of-the-art engineering approaches to individual hazards or infrastructure. There is no doubt that improvements in the constructed environment can minimize damage and thus economic losses on a structure-by-structure basis, but aggregated data on economic damage, damage avoided, residual damage, and a wide range of other losses are notoriously incomplete and unreliable. Furthermore, there is no measure of the extent to which reliance on engineering technology may have encouraged the process of ever-more-expensive development in hazardous places.

Nevertheless, carefully engineered buildings and infrastructure will be essential to the future disaster resiliency of all localities. The task will be to accurately assess all of the hazards involved and balance them against the benefits to be gained, according to a given community's preferences. Localities will need to reach their own decisions about the level of hazard they believe is appropriate to their situation and balance that with the other components of sustainability—the quality of life they want, the kind of environment they want future generations to have, and so forth. Part of that decisionmaking process will be to determine the extent to which an engineering solution will reduce future hazard, for how long, against what magnitude of extreme event, at what economic cost, and at what environmental price. This "total risk management framework" will be complex, unique to each locality, and probably changing continuously.

Engineering codes, standards, and practice have evolved and been promulgated in the United States for all natural hazards. For buildings and lifelines these codes vary widely based on local perceptions of benefits, costs, and risks. Affordability is a key factor in whether codes and standards are changed. Thus, the establishment and enforcement of codes and standards are a problem of social choice and are not based exclusively on technical feasibility. Buildings are complex combinations of the basic foundation and structure, plumbing, electrical, heating, ventilation, air conditioning, and ancillary systems. The structural aspects of a building consider earthquake, high-wind, flooding, and related disaster loads.

Infrastructure (sometimes called "lifelines") includes all of the structural components that provide connections among human developments, such as transportation networks, communications networks, electric power lines, gas lines, water supply systems, hazardous materials storage facilities, and drainage and water treatment systems. For infrastructure the engineer integrates all of the hazard impacts as part of the design and management. A recent report on the impacts of the 1994 Northridge earthquake presents a good summary of the impacts of this event on critical lifelines (Schiff, 1995).

Earthquakes

Historically and currently, life safety is the first goal of the seismic resistant design of structures. Reducing the loss of function of a structure or preserving property are secondary important goals. Since the first assessment was completed, observations of the effects of earthquakes have revealed that much is not yet known about the effects of earthquakes. Most estimates of ground motions are uncertain by a factor of two or more. Earthquake magnitude, distance from the source of energy release, location of a structure relative to the fault, and site response are each important factors in determining impacts. Earthquakes each have their own characteristics, e.g., variations in frequency content, accelerations, and duration. Building performance varies widely due to these differences; consequently, performance of a structure in one earthquake does not predict performance in another earthquake (California Seismic Safety Commission, 1994).

The major factors that influence earthquake-induced damage to structures include the characteristics of the site, the structural system itself, and its configuration relative to the site. These factors have as much influence on damage as do the size of an earthquake or distance to a fault. Most of what is known about the performance of structures in earthquake results from actual experience.

A related but separate issue is damage to the nonstructural elements of buildings since such damage impairs functionality. In fact, most life loss and injury in earthquakes is caused by the failure on building contents—for example, toppled bookcases and furniture, broken or fallen pipes, and collapsed suspended ceilings. Building contents frequently fail in earthquakes even when the building that houses them performs well.

Another important issue is the mitigation of the effects of earthquake hazards on lifeline systems. Most current efforts to mitigate impacts seek to design for operational flexibility in the face of lifeline service interruption; for example, storing water, electrical generators, and wireless communication devices. Lifeline replacement or constructing redundant lifelines are often too expensive to implement in most of the nation's communities; although noteable exceptions exist, for example, Pacific Gas and Electric Company's lifeline replacement program in northern California.

Significant current and future shifts in the nation's approach to earthquake engineering provide a clear link to sustainable hazards mitigation. These shifts include performance based seismic design which enable building owner decisions about the level of seismic resistance put into a building; for example, an owner can choose to go beyond life safety to include damage control to maintain operations after an earthquake. A second example is the clear call for increased laboratory testing of both structural and nonstructural building elements.

Floods

Flood control is a classic case of the evolution of engineering design standards in an area where a wide variety of structural and nonstructural controls have been used over a long period of time. Engineers have long recognized that flood protection cannot be provided for every conceivable flood. The methods previously used made some gross assumptions as to the uncertainty in hydrological and hydraulic calculations, the level of acceptable risk, and the benefits derived from flood control. The U.S. Army Corps of Engineers recently developed new methods for evaluating uncertainty in flood control evaluations (National Research Council, 1995). The methods explicitly estimate uncertainty in the hydrology, hydraulics, and economics of a planning study. The new methods bring these uncertainties to the forefront of the process used to evaluate various forms of flood control. Local decisions and analysis may now take into consideration uncertainty in engineering calculations, uncertainty in estimations of naturally occurring random events (i.e., precipitation, runoff), and future economic uncertainty.

Although these new methods do not yet include a cross-hazard holistic approach to decisionmaking, they may be a step in the right direction by bringing previously "assumed away" uncertainty into the decisionmakers' arena. In addition, very different flood control solutions may be weighed against each other in a consistent manner. For example, an improved levee system could be compared to a floodproofing program, where the probability of net benefits for each project could be weighed.

Dams and associated reservoirs serve multiple purposes, but they also increase hazards downstream if they fail. Engineers have developed sophisticated methods to evaluate alternative failure mechanisms for dams, including foundation conditions, materials, and initiating mechanisms such as floods, earthquakes, and landslides. Dam safety evaluations use relatively sophisticated methods of risk analysis, including comparing risks across hazards.

Hurricanes

Hurricanes can cause significant loss of life and economic damage. In the case of Hurricane Hugo, which hit the coast of South Carolina in September 1989, engineers had been recommending for 20 years that buildings and other structures be designed to withstand hurricane conditions with recurrence intervals of 50 to 100 years. Even though Hurricane Hugo was not that severe an event, the area suffered major damage. The major impact on lifelines was the failure of the electric supply system, which collapsed in winds of only 70 mph. Only 23 percent of the residents in the Charleston area had power eight days after the hurricane, and it was two to three weeks before power was restored in some rural areas. The failure of the power system caused problems with other infrastructure. Work by the telephone companies to move utilities underground had improved the reliability of the telephone system. Properly designed structures fared very well.

In contrast, Hurricane Andrew in 1992 caused relatively little damage to lifelines, although above-ground utilities, especially electric power lines, performed poorly. The major damage was to residential structures from the failure of roofing materials, doors, and windows, which led to weather penetration and significant damage. Nearly 100 percent of the manufactured housing was destroyed. Overall, the vast majority of monetary damage to buildings was due to water penetrating the buildings, not structural failure.

Winds

During the past 25 years, organizations around the world have documented the performance of buildings in wind storms on a comprehensive and systematic basis. Professional and academic institutions in the United States that documented wind damage include the American Association for Wind Engineering (formerly the Wind Engineering Research Council), the Institute for Business and Home Safety, Texas Tech University, Clemson University, Texas A&M University, and the National Research Council. This information provides insights into the actual performance of buildings in wind storms. The general observations on building performance and identification of problem areas made below are based on these field investigations.

The structural integrity of contemporary high-rise buildings (six stories and higher) should be maintained in wind storms. However, roofs and walls are susceptible to damage. Industrial buildings with frames of heavy steel and reinforced concrete generally maintain structural integrity in wind storms. But overhead doors, windows, and light-weight metal wall or roof panels are susceptible to failure.

Low-rise commercial buildings such as retail stores, schools, warehouses, and motels sustain extensive damage in severe wind storms. The level of damage depends on maintaining the integrity of the structure. If the structural frame collapses, there is total loss. Light steel frames, unreinforced concrete block walls, and timber frames, which are expected to maintain structural integrity, sometimes fail. The result is collapsed roofs and walls and possibly total collapse of the building, perhaps causing death or injuries to any occupants.

Single- and multiple-family (up to four units) dwellings, which are generally not engineered, experience a large amount of damage in wind storms. If not properly anchored to the walls, the roof lifts up, resulting in extensive damage, including subsequent collapse of walls. Residential buildings generally have several intersecting walls that create a small interior space; this interior space provides a high degree of protection to the occupants even when the rest of the building collapses.

The first national consensus standard that contained wind load provisions was American National Standards Institute A58.1-1955. There were very few changes over the years. The 1982 version of ANSI A58.1 contained a new basic design wind speed map and attempted to clean up the ambiguities and simplify the procedures of prior versions. Model building codes began to acknowledge the ANSI A58.1 standard and accept its provisions. The next revision came in 1988, when the American Society of Civil Engineers took over maintenance of the standard and it became known as ASCE 7-1988.

After Hurricane Andrew in 1992, code bodies and municipalities looked for more rigorous codes, and ASCE 7-88 was recognized as such. Finally, ASCE 7-95 was published in 1995. It contains a new basic wind speed map based on three-second peak gust wind speed rather than the antiquated fastest-mile wind. Model building codes are in the process of adopting ASCE 7-1995. The ASCE 7-95 code is based on extensive full-scale and wind tunnel studies of wind effects on structures. The standard recognizes the high localized pressures at eaves, ridges, and wall and roof corners. It accounts for internal pressure when there are openings in the building.

Landslides, Land Subsidence, and Expansive Soils

The main option for prevention of landslide damage through engineering is careful site preparation in the form of grading and attention to drainage, followed by improved construction practices to minimize instability of the slope. Some engineering solutions are available to minimize subsidence damage to structures and lifelines, including use of geofabrics and earth reinforcement for slabs and roads, proper log grading, and foundation and utility supports. Mines and other subsurface cavities can sometimes be backfilled. Engineering mitigation options for expansive soils include removing them, applying heavy loads to offset swelling pressure, preventing access of water, prewetting, and stabilization with chemicals.

Snow

Mitigation of the snow load hazard for new building construction is fairly straightforward. It involves the design or selection of structural members having adequate strength to resist the roof snow loads given in building codes and load standards. Use of roof loads intended to have a mean recurrence interval of about 50 years, in combination with structural safety factors, is thought to result in structures with an acceptably low probability of structural distress or collapse.

Mitigation of the snow load hazard for existing construction is more complicated. Assuming that the initial structural design was proper, problems can arise from deterioration in structural capacity or larger-than-expected loads. Currently, there are no commonly recognized government requirements for periodic inspection or upgrades of privately owned facilities. As a result, mitigation of existing structures is typically owner driven. For a roof it can take the form of structural strengthening of vulnerable elements or other more innovative approaches, particularly for drifts.

CONCLUSION

All of the mitigation tools described in this chapter are essential for a future that embraces sustainable hazards mitigation. The challenge for future researchers and practitioners will be to combine these tools in the most effective and economical ways to achieve community or regional goals of sustainability. Interestingly, the tools for sustainable hazards mitigation reviewed in this chapter are the same ones that have always been available for natural hazards management. What is needed is a shift in emphasis. The sustainability approach to hazards calls for increased use of wise, long-term land-use approaches, enhanced production and use of long-term hazard forecasts in community decisionmaking, insurance as a vehicle to foster mitigation efforts through location decisions and construction practices, and engineered approaches and building codes that go beyond life safety toward protecting the functionality of structures that localities choose to locate in harm's way.

CHAPTER SEVEN

Preparedness, Response, and Recovery

SUSTAINABLE HAZARDS MITIGATION will not eliminate the need for emergency preparedness and disaster response to deal with the physical destruction, losses, and human suffering imposed by disasters. In addition, there is a great opportunity to make progress toward sustainable hazards mitigation by setting policies and practices for recovering from disasters. A large body of research has been carried out on predisaster response planning and postdisaster response since the 1975 assessment (White and Haas, 1975), but it has not always produced consistent findings. Progress has been made in describing and analyzing preparedness and response activities, synthesizing what is known, developing a new theoretical approach, using new research methods, and applying knowledge.

DISASTERS

Considerable change took place in theorizing about the characteristics of hazards and disasters over the more than two decades since the first assessment. These theoretical changes have influenced the research that was performed and the knowledge accumulated since 1975, and they have begun to redirect national policy and find their ways into response and preparedness practices in the nation's local communities.

When the first assessment ended, most people agreed with the initial definition of disaster developed by Charles Fritz. He defined disaster some 35 years ago as:

> . . . an event, concentrated in time and space, in which a society, or a relatively self-sufficient subdivision of a society, undergoes severe danger and incurs such losses to its members and physical appurtenances that the social structure is disrupted and the fulfillment of all or some of the essential functions of the society is prevented. (Fritz, 1961, p. 655)

This functionalist viewpoint directed research and strongly influenced national policy. For example, it still underlies national logic about when affected communities can seek assistance from states and when states, in turn, can seek federal aid.

But today opinions about what constitutes a disaster diverge. Latter-day theories include social constructionism, postmodernism, and conflict-based and political economy theories (see Kreps and Drabek, 1996; Horlick-Jones, 1995; and Hewitt, 1983, respectively). Examples of this variation can be seen in a special issue of the *International Journal of Mass Emergencies and Disasters* (1995), which was devoted to discussions of the concept. In that issue, C. Gilbert views disaster as an analog for war, as social vulnerability, and as a condition that engenders uncertainty. W. Dombrowsky argues that disasters occur because the consequences of human activities are not well understood, frequently interact and conflict with technological and natural processes, and are on a collision course with nature. G. Kreps takes the position that Fritz's definition should be retained with the modification that disasters are social constructions; that is, disasters do not exist in and of themselves but are the products of how people agree to define them. B. Porfiriev defines disaster as a breaking of the routines of social life in such a way that extraordinary measures are needed for survival. Horlick-Jones argues that disasters originate in the conditions of modern society and are disruptions in cultural expectations that cause a loss of faith in the institutions that are supposed to keep hazards under control. Hewitt criticizes mainstream approaches for focusing on the physical characteristics of disasters because that tends to locate the source of the disaster outside society rather than within it.

The largest impact on preparedness and response research in the

United States has been made by the social constructionist perspective. It views disasters as socially produced through the formation of a common and shared definition. Hence, events like the Three Mile Island accident can be socially defined as a disaster even though there were few physical impacts, and floods that in earlier times might have been considered acts of God are now considered the responsibility of the U.S. Army Corps of Engineers.

Today theorists do not argue about the existence of earthquakes, floods, or tornadoes. Instead, they point to the importance of exploring the activities engaged in by groups as they try to place disaster-related problems on the public agenda and elicit particular kinds of governmental and institutional responses. Today, the properties of disasters are not seen as inherent in the phenomena themselves but rather as products of a social definition that is broadly agreed on by members of society.

The current trend in the United States is toward a synthesis of these two theoretical perspectives, and this has influenced research and policy. For example, research questions now relate to how social processes impact hazards mitigation; how government agencies have acted to distribute earthquake hazard information in ways that could "reconstruct" public attitudes and actions; and how the Federal Emergency Management Agency's National Mitigation Strategy seeks to alter the socially constructed salience of hazards and hazards mitigation among the nation's public and to change national culture to make people more instinctively aware of hazards and ways to avoid them.

Technological Disasters

Technological hazards, emergencies, and disasters are increasing more rapidly than natural disasters. Part of this increase may be due to improvements in the detection and measurement of technological hazards and part may be from increased vulnerabilities that sometimes result from the introduction of new technologies and increased population growth and concentration.

Since the nation's first assessment (White and Haas, 1975), a major ongoing debate has developed on the issue of whether natural and technological disaster agents differ in terms of the preparedness and response behaviors they stimulate. The points of view in this debate can be grouped into four categories, as follows:

1. *Technological and natural disasters are distinct.* One body of research suggests that disasters involving technological agents are distinct in several ways. Some studies suggest that technological disaster agents produce responses in the public that differ from those in natural disasters; for example, compliance with warnings is generally high with technological disasters. Technological disasters are thought to result in more serious or longstanding mental health problems.

Natural disasters are generally characterized as "consensus" crises that are accompanied by heightened community cohesiveness and morale. In contrast, technological hazards are thought to result in heightened levels of community conflict. Some of the other distinctions claimed are that the two pose different managerial problems; natural hazards and disasters, some say, are familiar while technological hazards are often unfamiliar; and there is only limited potential for avoiding natural hazards.

2. *The natural/technological distinction is not important.* A second position in the debate downplays the distinction. Some analysts suggest that other underlying dimensions do a better job of explaining differential responses to natural and technological hazards, such as familiarity, speed of onset, length of the warning period, and scope of the impact. For others the key lies in the social and social-psychological processes that apply when people are involved. Thus, natural and technological agents that resemble one another along particular dimensions will produce similar kinds of responses and challenges.

3. *Disasters as social constructions.* According to this view, people respond to phenomena in terms of the meanings they assign to them. By showing how views of disasters are shaped by social, cultural, and institutional practices, a social constructionist approach undermines the natural/technological distinction. For example, Stalling's (1995) analysis of the earthquake problem argues that the earthquake threat is socially constructed, the product of promotion and claims making by a group he terms the "earthquake establishment." Stallings is not, of course, arguing that earthquakes are not real; rather, his study documents the ways in which organized social actors frame the earthquake problem as a putative threat.

4. *Human agency as the unifying factor.* Another recently developed point of view is that the natural/technological distinction is an artificial

one because human agency is the key factor in all disaster events. In other words, rainfall may have been the trigger for the flooding, but the flood disaster was socially generated because people placed themselves at risk by living in a floodplain.

Recovery from Technological Disasters

Recovery from technological disasters is thought to present different concerns than recovery from natural events, but little research exists on the topic. Most significant U.S. technological disasters have not had a major or lasting impact on U.S. urban areas or populations. To date, most impacts from events such as the Exxon Valdez oil spill in 1989 have instead had rural and ecological impacts. There is little, if any, literature that describes state or federal efforts to anticipate long-range recovery issues, lasting impacts, or estimates of resources needed to deal with long-term recovery from a major toxic chemical or hazardous materials incident. No government agencies, at any level, have plans for long-term recovery from technological hazards.

Management Issues

Technological hazards pose some management problems that differ from those of natural hazards, owing to factors like the relative newness of some technological hazards, the lack of accumulated experience with control or coping measures, broader opportunities for control intervention, greater susceptibility of technological hazards to being "solved," and the need to simultaneously enlarge benefits and reduce risks in judging the tolerability of technological hazards. Also, unlike natural hazards, technological disasters result in a greater reliance on governmental authority and a reduced use of community and family social networks. Understanding differences like these is important to the study and development of effective ways to manage technological hazards.

Technological hazards, especially those associated with new technologies or those that are imposed, are assumed by many people to be amenable to "fixes" that will substantially reduce risk. Managing technological hazards requires the dual goals of enlarging social benefits and reducing hazards and accompanying actions to reduce inequities where the people who benefit from the technology are not the same people bearing the risk. Major social issues of public versus private responsibility for mitigation are raised with technological hazards. What incentives are most effective in encouraging private interests to reduce risks from technologies under their control? Can the private sector be persuaded to assume responsibility (beyond that demanded by regulation) for risk reduction that will benefit society?

Problems of Recovery

Despite the fact that there has been little research on recovery from technological disasters in the United States, some current work is pointing in potentially useful directions. In examining technological disasters from the past 40 years—such as the Chernobyl, Bhopal, and Exxon Valdez events—one study noted that they differed from natural ones in a lack of a sense of urgency for rebuilding, a priority for relocation versus rebuilding in place, more elaborate conflict between victims and nonvictims, and a delayed response by governmental authorities. Of the nine cases assessed, only one resulted in a more or less successful recovery (see Mitchell, 1996).

Long-Term Impacts of Disasters

A key issue in considering research on and policies for recovery and reconstruction is whether increased damage from disasters (see Chapter 4) translates into longer-term impacts. If states or communities prove to be resilient, with economies returning to predisaster conditions within a few months to a year, the case for government assistance is weakened. However, if the longer-term effects are substantial, there is a rationale for a strong governmental role in the future as losses increase.

The question of whether disasters have long-term economic effects was the subject of vigorous debate in the early 1980s. The general finding from one set of studies was that most disasters affect relatively small proportions of communities and that those communities as a whole tended to bounce back quickly with available forms of assistance (see Friesma et al., 1979). But in studying losses to public facilities after disasters, another researcher concluded that "while for many governments disasters are, in fact, manageable financially, a few governments—about one in 10 of those with losses—experience truly catastrophic damage to public property" (Burby et al., 1991, p. 46).

Another study shows that the issue is more complicated and that a key question is whether or not a certain type and magnitude of disaster is anticipated in the community. If so, the disaster, when it occurs, would have no long-term economic impacts. But if disasters of a different type, larger size, or greater frequency occur, local economic impacts can be expected (see Yezer and Rubin, 1987).

In short, there is accumulated evidence that many smaller disasters have little or no long-term economic impacts on communities or most individuals. The latter is due to a large extent to the availability of insurance and other forms of assistance. Yet these are relatively narrow findings that fail to take into account several unresolved issues. The most important is the limited understanding of the longer-term effects of large events, such as an urban earthquake, for local economies, social patterns, and key sectors of the economy.

DISASTER PREPAREDNESS

The purpose of preparedness is to anticipate problems in disasters so that ways can be devised to address the problems effectively and so that the resources needed for an effective response are in place beforehand. Preparedness includes such activities as formulating, testing, and exercising disaster plans; providing training for disaster responders and the general public; and communicating with the public and others about disaster vulnerability and what to do to reduce it. Since the first assessment, there has been an increasing emphasis in the nation on preparedness for chemical, nuclear, and other technological hazards. This was driven by heightened awareness as a result of events at Three Mile Island and Bhopal and from new legislation such as the Superfund Amendment and Reauthorization Act. Preparedness activities can be analyzed at various levels: families and households, organizations, communities, and states and nations. Factors that influence preparedness in each of these levels are discussed below.

Households

One of the most important contributions of the research conducted over the past 20 years has been to highlight the importance of socioeconomic factors in household preparedness. Other things being equal, households with higher socioeconomic status and nonminorities are better prepared than others, but even those that do prepare are doing relatively little. Purchasing insurance, making structural changes to the home, assembling first aid kits, storing food and water, rearranging furniture and the like, and making a household disaster plan are activities that some of the "prepared" households in one recent survey had done. But many people apparently had taken no action at all, even though they said they did expect a hazard to occur. Although some of the factors that affect preparedness are known, there is still no thorough understanding of the social-psychological processes involved in making the decision. In other words, researchers know who prepares but not why (see Mileti and Fitzpatrick, 1993).

A good deal also has been learned in the past two decades about how public risk information and communication can overcome obstacles to foster significant amounts of household preparedness. Public awareness campaigns designed to improve household preparedness have generally been undertaken in the context of growing awareness about a near-term threat, like the probability of an earthquake on the San Andreas fault at Parkfield. Less is known about what types of incentives will motivate people to increase and sustain preparedness efforts during periods of relative normalcy.

Finally, it should be emphasized that, although the volume of information on household disaster preparedness has grown tremendously, large gaps still exist, the research cannot be considered cumulative, and only some findings have been replicated. The fact that much of the work undertaken has been done in particular kinds of hazard contexts, such as situations involving a short-term likelihood of damage (e.g., the danger of earthquake aftershocks) limits the applicability of what has been learned.

Organizations

Like that on households, knowledge about organizational preparedness and the factors that encourage it is still quite uneven. Considerably more is known about preparedness among public-sector organizations than others, but even that is far from comprehensive. Furthermore, there has been a tendency to focus on organizational preparedness only for specific kinds of hazards.

Local Emergency Management Agencies

There is agreement that preparedness among local emergency management agencies in the nation has improved significantly. Over time more community organizations appear to have become interested, and planning is becoming more integrated. As in the 1960s and 1970s, local emergency management agencies remain diverse in their organization and operations, including jurisdictions and responsibilities, relationships with other emergency-relevant organizations, and resources available. They are, however, well adapted to their local situations. The International City/County Management Association's 1982 survey of more than 6,000 local and county governmental units found considerable structural variation and lack of standardization.

Four general factors have been suggested as contributing to successful local emergency management agencies: the existence of persistent hazards; integration of the emergency management office into the day-to-day activities and structure of local government; extensive relationships with other community organizations; and concrete outputs to the community, such as the maintenance of an emergency operations center (see Wenger et al., 1986).

Despite the improvement, problems remain. There is still a tendency to base plans on disaster myths rather than on accurate knowledge, to plan in isolation, to emphasize command and control, to focus on written products (like plans) rather than on the process, to succumb to overconfidence based on successful response to routine emergencies, and to accept the low priority of disaster planning as an excuse for inaction. Furthermore, preparedness has been found to be fragmented rather than integrated across different local organizations and sectors; as a result, planning for disasters tends to be done in isolation.

Fire, Police, and Medical

In the past 20 years almost nothing has been learned about police and fire department disaster preparedness. There is at most a general idea of how they plan and for what types of tasks and events. It has been found that police departments, especially smaller ones, tend not to devote much internal energy to disaster planning. When they do plan, police agencies tend to do so in isolation from other community organizations; few have adopted an interorganizational approach to disasters. The police appear to believe that disasters can be handled through the expansion of everyday emergency procedures (see Wenger et al., 1989).

Fire departments have improved their preparedness levels and expanded their disaster- and crisis-related tasks beyond fire fighting. In particular, they tend to be involved in planning for the provision of emergency medical services and for responding to hazardous materials emergencies. Nevertheless, like police departments, fire departments show a tendency to only plan internally (see Wenger et al., 1989).

Most of what has been learned in the past 20 years concerning the preparedness of emergency medical service (EMS) providers comes from studies conducted in the mid- to late 1970s. Those studies showed that, like fire and police units, EMS providers tend to plan in isolation and believe that expansion of everyday activities can cover a disaster situation. There is evidence that hospitals and health care organizations are not prepared to advise people about or to treat victims of chemical hazards and disasters.

No research has been conducted on how the changes in health care are affecting planning for disasters, but since disasters never were a major priority for most EMS organizations, it can probably be assumed that they have moved down on the agenda.

Agencies Without Crisis Missions

Three factors have been consistently identified in the research as having positive influence on disaster preparedness among nonemergency organizations. First, larger organizations are more likely to have both greater resources and a more urgent need for strategic planning and thus a greater concern for disaster preparedness. Second, the level of perceived risk of organizational or department managers is positively correlated with preparedness. Third, the extent to which managers report seeking information about environmental hazards is positively correlated with organizational preparedness.

The Private Sector

Until relatively recently, business preparedness was virtually never investigated by researchers. The research that does exist indicates that private firms are less than enthusiastic about disaster preparedness, even in hazard-prone areas. The strongest predictor of preparedness levels among businesses is size, followed by previous disaster experience, and owning rather than leasing business property.

Communities

Numerous studies have shown that local support for disaster preparedness is low in most communities and that relatively few resources are allocated to disaster preparedness and response. This low priority of disasters tends to occur because disasters are infrequent in any given locality, responders tend to overgeneralize from their experiences with routine emergencies, and nonspecialists tend to either underestimate the magnitude of disaster demands (resulting in unrealistic optimism) or grossly overestimate them (resulting in fatalism).

The Emergency Planning and Community Right to Know Act, Title III of the Superfund Amendment and Reauthorization Act of 1986 (SARA), mandated the creation of local emergency planning committees (LEPCs) nationwide to, among other things, prepare for hazardous chemical releases. At the time the law was enacted, communities were not well prepared for major emergencies involving hazardous chemicals. A number of studies on LEPCs themselves followed, but there is still not a great deal known about their implementation or effectiveness.

State and National Preparedness

States possess broad authorities and play a key role in disaster preparedness and response, both supporting local jurisdictions and coordinating with the federal government on a wide range of disaster-related tasks. Federal resources usually cannot be mobilized in a disaster situation without a formal request from the governor, and states have a number of their own resources at their disposal for use in emergencies, including the National Guard. States are required to develop their own disaster plans, and they typically also play a role in training local emergency responders. States have responsibilities for environmental protection and the delivery of emergency medical services, and state emergency management duties have broadened in recent years as a result of legislation like SARA Title III, which requires states to coordinate the chemical emergency preparedness activities of local LEPCs.

In view of the important roles that states play in the management of hazards and disasters, the vanishingly small amount of research that is focused on state-level disaster preparedness activities is surprising. In the mid-1970s the National Governors' Association compiled detailed information from 57 states, commonwealths, and territories on their emergency preparedness and response activities. In the nearly 20 years since the association's report, there have been only a few scattered studies on preparedness measures undertaken by states, and their scope has been limited. Few comprehensive studies have been done. What states do undoubtedly makes a difference at the local level; however, without research that takes an in-depth look at what states and localities are actually doing, researchers can conclude little about their role in the preparedness process.

The picture is scarcely better at the national level. Much of the knowledge in hand about federal government preparedness comes from detailed case studies that either focus on the federal government at a particular point in time or assess changes in federal policies and programs that have taken place over time.

In the United States federal disaster preparedness evolved out of an earlier concern for civil defense. At times, this emphasis made implementing preparedness measures difficult. For example, an important factor in local resistance to federally directed "crisis relocation planning" for disasters was the program's relationship to nuclear war readiness (see May and Williams, 1986).

National-level preparedness initiatives tend also to be shaped by dramatic events. For example, the Three Mile Island nuclear accident stimulated federal action to encourage extensive evacuation planning for areas around nuclear power plants, and the Bhopal disaster was a major factor influencing the content of SARA Title III, which mandated local emergency planning. Similarly, many provisions in the Oil Pollution Act of 1990 were a direct response to the Exxon Valdez oil spill. The federal response plan had already been developed before Hurricane Andrew in 1992, but the delayed and uncoordinated response to that disaster prompted calls for improved response planning.

One key message in the research literature is that federal preparedness is influenced and constrained not only by institutional power differentials but also by the nature of the intergovernmental system itself—the nature of federalism; the complexity of agencies, responsibilities, and legislation; and the difficulty of effective interagency coordination.

RESPONSE

Disaster response activities include emergency sheltering, search and rescue, care of the injured, fire fighting, damage assessment, and other emergency measures. Disaster responders must also cope with response-generated demands such as the need for coordination, communications, ongoing situation assessment, and resource mobilization during the emergency period.

The response period has been the most studied phase of disasters. In general, response research has a good deal in common with the research

U.S. Army infantry troops arrive in Bozeman, Montana. They were the first of 14,000 soldiers and Marines flown in to fight the 1988 fires in Yellowstone National Park. Photograph by Larry Mayer, courtesy of the *Billings Gazette.*

on preparedness. Conceptual frameworks, research designs, and the variables included in analyses range widely across studies, making generalizations difficult. Some topics have been studied extensively, while others have received little emphasis. And studies that focus on smaller human groups such as households are much more common than broad ones.

Emergency Shelter and Housing

Most knowledge about postdisaster sheltering and housing comes from research done in the past 15 years. Little is known about housing patterns across social classes, racial/ethnic groups, and family types. Postdisaster sheltering and housing encompass both physical and social processes and different phases as the disaster progresses into the recovery period.

Housing patterns after disasters are influenced by predisaster conditions, such as interorganizational mobilization and communications; preimpact community conflict, resource and power differentials, and both victim and community sociodemographic characteristics. Predisaster social ties influence where people go when they flee their homes; those who have small or weak social networks are more likely to use public shelters, while others go to family members and friends. Preexisting social inequities, including differences in income and household resources, home ownership, insurance, and access to affordable housing also have a significant impact on housing options after disasters (see Quarantelli, 1982).

Social Solidarity

The literature on U.S. disasters consistently shows that social solidarity remains strong in even the most trying of circumstances. There is no doubt that disasters engender prosocial, altruistic, and adaptive responses during the emergency period immediately after a disaster's impact.

Behavior in disaster situations is adaptive, and considerable continuity exists between pre- and postdisaster behavior patterns. Rather than being dazed and in shock, residents of disaster-stricken areas are active and willing to assist one another. Important response work typically is performed by community residents themselves. Volunteer activity increases at the time of disaster impact and remains widespread during the emergency period. It can emerge spontaneously or be institutionalized as part of an organization that takes on volunteers such as the American Red Cross. In addition, there are so-called permanent emergency volunteers who regularly get involved in response activities.

Group Emergence

New groups invariably form (or emerge) during and after disasters, usually in situations characterized by a lack of planning, ambiguity over legitimate authority, exceptionally large disaster tasks (e.g., search and rescue), a legitimizing social setting, a perceived threat, a supportive social climate, and the availability of certain nonmaterial resources. Political and social inequality may also drive emergence.

Emergent citizen groups are typically composed of a small active core, a larger supporting circle, and a still larger number of nominal supporters. They usually have few monetary resources, but volunteer time and commitment are more important than financing. Sometimes emergent groups will transform themselves after a disaster to address more general community needs and persist as an organization.

Organizational Response

Organizations responding in disaster situations face a number of challenges. They must mobilize; assess the nature of the emergency; prioritize goals, tactics, and resources; coordinate with other organizations and the public; and overcome operational impediments. All of these activities must be accomplished under conditions of uncertainty, urgency, limited control, and limited access to information. In the absence of prior interorganizational and community planning, each affected agency will tend to perform its disaster-related tasks in an autonomous, uncoordinated fashion. One of the challenges of disaster planning and management is to overcome the natural tendency organizations have to maintain their independence and autonomy and encourage them to have a broader interorganizational and communitywide focus (see Kreps, 1991).

Local Emergency Management Agencies

Early emergency response research pointed to the difficulties that local agencies had in actually managing the response in disaster situations. That situation evidently has improved over time, although the improvements have been uneven.

Local emergency management agencies are widely varied in their assigned responsibilities, in their relationships with other community organizations, in how they carry out their emergency-related tasks, and in the amount and kinds of crisis-relevant resources under their control. They have many different patterns of organization, ranging from those in which emergency management agencies are weak, isolated, or bypassed during the response period to those that are well institutionalized and embedded in a communitywide emergency management system.

The use of emergency operations centers (EOCs) in the management of disaster response operations has now become common. In general, EOCs work best when they are adequately staffed and supplied, have management and communications systems, survive the disaster impact, and have clearly delineated functions. Beyond that, little is known about what makes them effective.

Fire, Police, and Medical

Studies of response operations seldom focus specifically on these organizations. One study did note that structural alterations within police and fire departments are more likely when a disaster is extensive, where resource levels are low, and where there has been little prior planning; that decisionmaking becomes more diffuse during disasters than in nondisaster times; and that problems with communications and convergence are common. Fire departments were found to undergo fewer organizational changes during disasters and in general to have fewer problems in disaster situations than police departments. Both fire and police departments prefer a high degree of autonomy in their everyday operations, and these patterns carry over into disaster situations (see Wenger et al., 1989).

As is the case with the other crisis-relevant organizations, there is not a large body of work on how providers of emergency medical services (EMS) perform in disasters. As noted above, postdisaster search and rescue activities are typically performed by non-EMS personnel. The transportation of disaster victims to hospitals is almost invariably uncoordinated; there is usually an oversupply of EMS resources (especially transportation resources like ambulances) after a disaster; triage tends to be "informal, sporadic, and partial," and central coordination of emergency care activities is rare (see Quarantelli, 1983, p. 76).

The Private Sector

Until fairly recently, studies of the response of private-sector organizations in disaster situations were virtually nonexistent, and to date very few systematic studies have been done. Thus, little is known about how private organizations actually respond when faced with disaster-related demands. Existing studies tend to focus on particular types of organizations and rather narrow topics; small and nonrepresentative samples also limit their generalizability. Interviews with tourist industry executives indicate that they engaged in many of the same kinds of behaviors and decision processes as community residents in deciding what to do about disaster warnings; they tried to confirm warnings and consult with outside information sources.

The News Media

Interest in the news media in disasters is recent but increasing, driven both by large-scale disasters that became major media events and by a growing interest, particularly in the communications field, in studying disaster reporting. On the one hand, disasters are framed by news organizations in ways that can be misleading and especially oversimplified. The media can convey erroneous impressions about the magnitude and even the location of disaster damage, as occurred, for example, when San Francisco was characterized as virtually in ruins after the Loma Prieta earthquake, when in fact the city was only selectively damaged. To the extent they perpetuate myths about disaster behavior, the news media convey unrealistic impressions about disaster-related needs and problems, potentially leading both the public and decisionmakers to worry about the wrong things.

On the other hand, the news media also make a strong positive contribution in disaster situations. Effective warnings broadcast through the media are widely credited with reducing casualties from hurricanes, tornadoes, and floods. By reporting extensively on disasters and the damage they create, the media can help speed up assistance to disaster-stricken areas, and postdisaster reporting can provide reassurance to people who are concerned about the well-being of their loved ones. Good science reporting can educate the public about hazards, and in-depth stories can help provide the basis for informed hazard reduction decisions.

Multiorganizational Response

Networks of organizations operating during disaster response differ in terms of which organizations are central and how they are structured; the mix of organizations involved; lack of cohesiveness, particularly interorganizational communications; ambiguity of authority; and poor utilization of special resources (see Drabek, 1986).

Interorganizational and community response to chemical plant explosions and airborne chemical releases has been found to be marked by poor coordination among responding organizations and between the public and private sectors. Problems with obtaining and disseminating accurate information were common, and the involvement of extra-community organizations further complicated response. Situational factors (time of day, day of the week, whether an emergency occurred during daylight or nighttime) are important in either facilitating or hindering response.

Community Response

The disaster-stricken community has been described as altruistic, therapeutic, consensus oriented, and adaptive. Early studies documented enhanced community solidarity and morale after disaster impact, suspension of predisaster conflicts, a leveling of status, increased community participation, and shifting in community priorities to emphasize central tasks, such as the protection of human life.

Little of the research conducted over the past 20 years contradicts this, although some exceptions have been found. Disasters can become occasions for organized resistance against established institutional structures and bureaucratic procedures. After the Loma Prieta earthquake, for example, preexisting community conflicts reemerged rapidly after the disaster. Ethnic solidarity was a major factor in that resistance (see Simile, 1995). Certain technological disasters and chronic technological hazards have engendered conflict rather than solidarity.

State Response

Currently very little is known about state-level disaster response. One study of intergovernmental coordination found great variability in response for different disasters. Response activities analyzed in that study tended to be judged more positively when they were initiated and managed by lower levels of government. Also, both actual and perceived governmental effectiveness were related to disaster magnitude, the extent to which governmental agencies were prepared, and the public's capacity to cope with disaster impacts (see Schneider, 1995).

FACTORS THAT INFLUENCE PREPAREDNESS AND RESPONSE

Broader social, political, economic, cultural, and institutional contexts shape disaster preparedness and response, just as they influence the adoption of other mitigation measures as discussed in Chapter 5. At the personal and household levels, ethnic and minority status, gender, language, socioeconomic status, social attachments and relationships, economic resources, age, and physical capacity all have an impact on the propensity of people to take preparedness actions, to evacuate, and to take other mitigation measures. In addition, people use a wide variety of decisionmaking processes, not all rational.

Household preparedness activities are more likely to be undertaken by those who are routinely most attentive to the news media (i.e., those who are educated, female, and white); are more concerned about other types of social and environmental threats; have personally experienced disaster damage; are responsible for the safety of school-age children; are linked with the community through long-term residence, home ownership, or high levels of social involvement; have received some sort of disaster education; and can afford to take the steps necessary to get prepared.

For organizations, governments, and people in general, mandates and legal incentives can in some instances induce preparedness, proper response, and other actions. For example, it is unlikely that formal disaster plans would have become almost universal at the local level if they had not been required. Similarly, it is clear that changes in the regulatory environment after the Three Mile Island accident did spur nuclear facilities and nearby communities to expand their planning efforts.

The same factors that influence preparedness at the household level also exist for organizations: hazards have low salience for most organizations except when there is an imminent threat; disaster-related problems must compete with more pressing concerns; organizations in financial difficulty will downplay preparedness unless it is essential; the resources necessary to deal with it may not be adequate; there is a lack of clear and measurable performance objectives; public and official support is inadequate; and the local emergency management organization lacks expertise. Three additional factors help influence preparedness and response: the governmental context, disaster experience, and the progress in professionalization of emergency management, discussed below.

The Intergovernmental and Policy Context

The organization, the effectiveness, and in particular the diversity of preparedness and response efforts in the United States are in large measure a consequence of the structure of hazards-related management policy, which is in turn embedded in the intergovernmental system. A number of U.S. studies have pointed out the difficulties inherent in conducting hazards reduction policy in a social context in which responsibility for different aspects of a problem is diffused among governmental levels and agencies and in which both accountability and the ability to implement policy are weak. In addition, the legislative authority is diffuse and often adopted in the wake of a particularly compelling disaster. There is debate (and often confusion) over the respective roles of governments, and they tend to change over time. Emergency management's own goal has expanded in the past 20 years, beyond the civil preparedness mode that had shaped it for decades.

In the United States disaster preparedness and response are primarily the responsibility of local governments. However, emergency management is typically not a priority at that level, and local capacity and financial resources are easily overwhelmed. Very little research exists on the ways in which this and other aspects of governmental structure and policies influence preparedness and response activities.

Disaster Experience

In general, research suggests that prior disaster experience is a major predictor of higher levels of preparedness and more effective response, both for households and organizations, largely because it leads to greater awareness of the consequences of disasters and the demands that disasters generate.

Professionalization

Since the nation's first assessment of natural hazards there has been a trend toward professionalization in emergency management. A generation ago the position of local civil defense director was not considered a full-time job, and the skills needed to perform it were not well defined. Since the 1970s, full-time local emergency managers have become more common, and their roles have broadened beyond civil defense and immediate postimpact emergency activities. With the notion of comprehensive emergency management, the role has expanded to include mitigation and recovery.

Today there is wide acceptance of the idea that managing disasters requires specialized knowledge, skills, and training. The process of professionalization has been accompanied by the formation of organizations and associations concerned with the training of and awarding credentials to emergency management specialists, the development of specialized publications, and the spread of professional meetings and training. For example, the Federal Emergency Management Agency's (FEMA) Emergency Management Institute provides instruction in emergency management for state and local officials, emergency managers, volunteer organization personnel, and practitioners in related fields. Each state emergency management office has a FEMA-funded training officer who coordinates federally funded training programs throughout the state.

More colleges and universities are offering emergency management courses and undergraduate degrees in emergency management. The National Coordinating Council on Emergency Management (now the International Association of Emergency Managers) has awarded certification in emergency management to over 600 people since 1993.

RECOVERY AND RECONSTRUCTION

Since the late 1970s, new models of recovery have been developed; long-term disaster impacts and the effects of government policy on community recovery have been examined; and, recently, the relationship between recovery and mitigation has been examined. It has been observed that sustainability may hold the key to incorporating mitigation into recovery and reconstruction.

The term "recovery" has been used interchangeably with reconstruction, restoration, rehabilitation, and postdisaster redevelopment. Regardless of the term used, the meaning has historically implied putting a disaster-stricken community back together. Early views of recovery almost exclusively saw it as reconstruction of the damaged physical environment. Its phases went all the way from temporary measures to restore community functions through replacement of capital stocks to predisaster levels and returning the appearance of the community to normal to the final phase, which involved promoting future economic growth and development. Researchers discovered that communities try to reestablish themselves in forms similar to predisaster patterns and that the resulting continuity and familiarity in postdisaster reconstruction enhance psychological recovery.

Recovery as a Process

Other researchers view disasters as opportunities to address long-term material problems in local housing and infrastructure. They recast reconstruction into a developmental process of reducing vulnerability and enhancing economic capability (see Anderson and Woodrow, 1989).

The contemporary perspective is that recovery is not just a physical outcome but a social process that encompasses decisionmaking about restoration and reconstruction activities. This perspective highlights how decisions are made, who is involved in making them, what consequences those decisions have on the community, and who benefits and who does not. The process approach also stresses the nature, components, and activities of related and interacting groups in a systemic process and the fact that different people experience recovery differently. Rather than viewing recovery as linear with "value-added" components, this approach views the recovery process as probabilistic and recursive. This perspective has shifted the emphasis to examining differential group involvement and away from cataloging reconstruction constraints (see Berke et al., 1993).

Researchers have demonstrated that there are patterns in the recovery process. Recovery is a set of actions that can be learned about and implemented deliberately (see Rubin, 1995). It is possible to anticipate some of the main concerns, problems, and issues during recovery and to plan accordingly (see "Planning for Recovery and Reconstruction" below). Local participation and initiative must be achieved. Some of the key deterrents to speedy recovery have been identified through research—namely, outside donor programs that exclude local involvement; poorly coordinated and conflicting demands from federal and state agency-assisted programs; staff who are poorly prepared to deal with aid recipients; top-down, inflexible, standardized approaches; and aid that does not meet the needs of the needy (see Berke et al., 1993).

Findings from Recovery Research

Research on recovery and reconstruction has yielded information about households and families, organizations and businesses, and communities as a whole.

Households

Most research has focused on family recovery and has sought to answer such questions as: What type of families are most disrupted by a disaster? What family types recover most quickly? What things account for different rates of family recovery? Different models of family recovery have been used, and this has led researchers to investigate different things. Most, however, have examined how recovery is affected by a family's socioeconomic status and other demographic characteristics, position in the life cycle, race or ethnicity, real property losses, employment loss,

loss of wage earner(s), the family's capacity (economic reserves, extended family support and assistance), and the use of extrafamilial assistance programs. Past studies also defined recovery differently: how successful victims were in regaining predisaster income levels, the relationship between magnitude of losses and available economic resources, whether families eventually returned to their original damaged home or a comparable one, and whether the victim family believed that it had recovered.

Despite the differences in the methods they used, researchers have agreed that the following factors affect family recovery. Note that many of these parallel the influences on preparedness and response, described later in this chapter, and on the adoption of mitigation discussed in Chapter 5.

Linkages to extended family are strengthened immediately after disasters, and this lasts well into recovery. Extended kin groups provide assistance to relatives. Socioeconomic status, race, ethnicity, and even gender are interrelated in complex and different ways. Ethnic and racial minority groups are typically disproportionately poorer, and disproportionately more vulnerable to disaster and to the negative impacts of long-term recovery. Poorer families have more difficulty recovering from disasters and also have the most trouble acquiring extrafamilial aid.

Rural victims were more likely to use their kin group as a source of emergency shelter than urban families. In rural areas, high-income victims had fewer losses than low-income families; but in urban areas, income seems to make no difference. Rural families are less likely to receive extrafamilial assistance than urban victims.

Businesses

Businesses have many of the same characteristics as households. For example, they vary in size, income, and age; they are physically housed in structures that are more or less vulnerable; and they differ in the resources they control. Some businesses are obviously less vulnerable to disaster and more capable of recovering than others. Businesses play vital community roles, but research to date has not documented the effects of business closures on family and community recovery. Obviously, the longer businesses are closed, the greater the economic strain on employees' families and the greater the impact on local government revenues. The character of the community may be altered as people leave to market, shop, bank and use recreational facilities elsewhere or, if, for example, their children go to schools at a greater distance from their homes.

Communities

There are many components of community recovery—residential, commercial, industrial, social, and lifelines—and there are various degrees of recovery. Some aspects of community life, such as tax revenues and property values, may take years to return to normal. Successful recovery efforts typically include strong local community participation and integration of the community into regional and national networks. This requires not only strong local governmental capacity but also a cohesive system of public, private, and volunteer groups that are integrated within the community. State and national governments can provide relief and recovery assistance to communities, but with assistance comes increased interaction among officials across government levels and increased state and federal intervention into local decisionmaking. The intergovernmental relations can be facilitated through strong local leaders, local government capacity to take action on its own behalf, and local officials who understand state and federal disaster relief programs (see Rubin, 1995).

Planning for Recovery and Reconstruction

With each disaster, new knowledge is gained about how to plan more effectively for recovery and reconstruction. But this information and experience have not been systematically collected nor synthesized into a coherent body of knowledge.

Regardless of the severity of a disaster or the level of assistance from higher levels of government, the primary responsibility for recovery ultimately falls on local governments, yet little information exists to guide local decisions. It is possible to educate and train public officials to cope effectively with recovery in their jurisdictions, but planning for recovery has been minimal in the United States. This is changing. For example, FEMA's Emergency Management Institute now offers courses on recovery, and some states now offer recovery training to local officials. Domestic and international case studies do exist, largely because of the efforts of such groups as the United Nations Disaster Relief Organization and the Office of Foreign Disaster Assistance of the U.S. Agency for International Development, which have assisted many nations in the task of rebuilding.

Postdisaster Environment

After disasters, critical policy choices emerge, forcing unwelcome decisions on local government about whether to rebuild quickly or safely. Postdisaster recovery and reconstruction planning and management commonly reflect an effort to balance certain ideal objectives with reality. Recovery is characterized by wanting to (1) rapidly return to normal, (2) increase safety, and (3) improve the community.

Time is by far the most compelling factor influencing local government recovery decisions, actions, and outcomes. Viewed theoretically, policy decisions might range from major acquisition of hazardous lands at one end to minor modifications in the construction code at the other. Viewed practically, real decisions are likely to be severely limited by economic pressure and pressure to decide quickly. The pressures to restore normalcy in response to victims' needs and desires are so strong that safety and community improvement goals—modifying land use, retrofitting damaged buildings, creating new parks, or widening existing streets—are often compromised or abandoned.

Preevent Planning

The notion of predisaster planning for postevent recovery is a relatively new and powerful concept. When further developed, tested, and evaluated, such knowledge may help many communities mitigate current hazards before a disaster and recover more quickly and safely afterward. Preevent planning organizes community processes for more timely and efficient postevent action, clarifies key recovery roles and responsibilities, identifies potential financing, minimizes duplicative or conflicting efforts, avoids repetition of other communities' mistakes, and achieves greater public safety and community improvement. Most importantly, it can help communities "think on their feet" and thus be flexible enough to adapt their postdisaster actions to the actual conditions.

Interest in preevent planning for postdisaster recovery and reconstruction has grown in the past decade. For example, local, state, and private entities in California have developed several training exercises, manuals, and guidelines for earthquake recovery. The Los Angeles Emergency Operations Organization's plan was generally helpful in recovering from the Northridge quake, although there were some problems in implementation (see Spangle Associates, 1997).

Plan Elements

To be effective, local recovery plans should have the following components:

Community involvement. In the absence of consensus, recovery can become politicized and foster conflict. The plan must have been fully discussed, agreed to, and accepted by the community before a disaster occurs.

Information. Plans are only as good as the information they contain on (1) the characteristics of the hazards and the areas likely to be affected; (2) the population's size, composition, and distribution; (3) the local economy; (4) the resources likely to be available after disaster; (5) the powers, programs, and responsibilities of local, state, and federal governments; (6) existing land-use patterns and building stock location and characteristics; and (7) local infrastructure.

Organization. Plans should anticipate the need for topic-specific groups or organizations. Having an official rebuilding/restoration organization, for example, in place immediately after a disaster helps ensure that wiser decisions will be made.

Procedures. Following standard political processes for making recovery decisions consumes time, which delays action. Construction procedures, permitting, and code review should be streamlined (but not relaxed) after disasters. If possible, recovery plans should incorporate a vision of hazard reduction after disasters through altered land-use patterns.

Damage evaluation. Plans should be made for the mobilization, deployment, and coordination of building damage inspection; standards for repair and reconstruction should be set in advance; and relocation should be an explicit option.

Finances. Information should be on hand about basic government and private programs for obtaining recovery, relocation, acquisition, and other funding. For example, it can be extremely expensive to alter land-use patterns after a disaster, and in major recent cases relocation occurred only when funds were provided by the federal government.

Usefulness of the Plan

Many normal planning procedures are suspended when disaster-stricken communities start to recover and reconstruct themselves. Multiple and potentially conflicting goals are being sought simultaneously but at a faster rate than normal. Extraordinary teamwork is required among various local government departments. Planners must shift from an otherwise slow, deliberative, rule-oriented procedure to one that is more flexible, free wheeling, and team oriented. For example, after the 1991 Oakland Hills, California, fire, planners emerged from the early recovery with a different perception of their roles, heightened awareness of the complexities of implementing hazard mitigation within a turbulent postevent milieu, and greater acceptance of team-oriented responsibilities. Planners interviewed after that event revealed a feeling that Oakland had been essentially unprepared for the recovery issues it faced. They thought that a recovery plan formulated before the disaster might have not only reduced the time needed to sort through the various policy decisions that instead had to be dealt with from scratch but also would have helped them anticipate the pressures they encountered (Topping, 1998).

Recovery Policy

Disaster recovery and reconstruction have been quite aptly described as a set of processes in search of a policy. There are various federal loan and other assistance programs to assist long-term recovery, and the provisions of the federal tax code that allow for exempting disaster losses provide a hidden form of longer-term financial assistance. Like many aspects of policymaking for disasters, these programs have been built up over the years in response to specific events and problems; they are not the result of an intentional policy for ensuring speedy and sensible recovery (see May, 1985).

The absence of a federal policy for recovery assistance is explained by several factors. In withholding appropriations for such purposes as part of the Disaster Relief Act Amendments of 1974, federal policymakers cited concern over the potential costs of funding recovery programs and their effectiveness. Some subsequent research failed to find any long-term economic impacts from most disasters, further undermining the case for federal funding for long-term recovery. Perhaps most importantly, as disaster costs have mounted, federal policymakers have been reluctant to open up new avenues for relief assistance.

Special federal grants or loans after disasters have been documented in most instances as important in hastening recovery. Federal funds for recovery and reconstruction tend to be a small but nonetheless important part of recovery financing. They can cause problems, however, if recovery decisions or actions are delayed in order to tailor them to the requirements of federal programs. It is also important to note that the economic conditions of a city have a lot to do with the pace of recovery.

The Role of Mitigation

The postdisaster period is an opportunity to upgrade the quality of construction to better resist subsequent events and begin to think through ways to mitigate future damage. Recovery processes can also be used as opportunities to advance programs already in place, such as urban renewal, traffic bottlenecks, architectural incompatibilities, and nonconforming uses. Since the early 1980s FEMA has required postdisaster mitigation planning. With amendments to the Disaster Act in 1988, the rules were modified to allow a greater percentage of disaster relief monies for funding mitigation programs. This provides an important link between recovery processes and disaster mitigation efforts.

This seemingly sensible approach to recovery and mitigation is not always easy to achieve, however. The findings from studies of the implementation of planning and reconstruction recommendations after the 1964 Alaska earthquake, of land-use planning after other earthquakes, and of local recovery in the early 1980s after several presidentially declared disasters, point to a strong bias on the part of decisionmakers toward maintaining the status quo. These sagas are largely ones of missed opportunities. Time after time, local leaders failed to take advantage of the recovery period to reshape their devastated communities in a way that would improve their ability to withstand future disasters. Part of the explanation lies in the complexities of recovery and the fact that compressed time periods make it difficult to systematically evaluate options. But the lack of clear goals at federal, state, or local levels, the complexity of acting in concert with multiple entities, and the absence of institutional capacity brought about by advance planning all help undermine recovery targeted toward mitigation or sustainability. In some cases, of course, dramatic changes in the postdisaster character of the area are nearly impossible to implement—as in the case of a damaging earthquake in a heavily urbanized area. The patterns of urban land use and extensive investment in infrastructure make it impractical to do much besides reconstruct within much the same framework, although perhaps with improved building-by-building construction.

At the same time, there are many instances in which mitigation measures were implemented during recovery. Some land-use changes were put in place after the 1964 Alaska earthquake, for example, leaving an "earthquake park" open space. Several small communities were relocated away from flood-prone areas after the 1993 Midwest floods, with significant federal financial assistance.

A case study of state and local action after Tropical Storm Alberto illustrates that localities can be receptive to state and federal actions if the locals do not have to incur the immediate costs of providing them. In that situation the state of Georgia and the governor very capably established and implemented a response to meet both the immediate needs of victims and the anticipated long-term recovery needs of the affected communities. Highlighting this was the hiring and assignment of building code inspectors to communities to guarantee that rebuilding would meet current flood standards (see Mittler, 1997).

Furthermore, after Hurricane Andrew, Florida finally enacted a law (Senate Bill No. 1858, Ch. 93-211) that established a dedicated funding source for emergency management, thereby providing the state with sufficient resources to develop its emergency management capability at both state and local levels. Under this law, emergency management is no longer subject to annual budget negotiations. The measure had been defeated annually since 1990.

These examples reinforce the importance of two factors in bringing about postdisaster improvements. One is the existence of a preexisting plan or ongoing process for reshaping a community. This provides a commitment to follow through and the necessary preexisting knowledge of potential options. A second factor is the availability of outside funding to help bring about the desired changes.

Future Directions for Recovery

Achieving patterns of rebuilding that generally keep people and property out of harm's way is increasingly viewed as an essential element of any disaster recovery program. Rebuilding that fails to acknowledge the location of high-hazard areas is not sustainable, nor is housing that is not built to withstand predictable physical forces. Indeed, disasters should be viewed as providing unique opportunities for change—not only to building local capability for recovery—but for long-term sustainable development as well.

Integrating sustainable development with disaster recovery will require some shifts in current thinking, land use, and policy. Some broad guidelines for developing recovery strategies that promote sustainable hazards mitigation are enumerated below, based partly on principles suggested by Mitchell (1992):

- Sustainable hazards mitigation calls for the adoption of a much longer time frame in recovery decisionmaking. Particularly foolhardy are short-term actions that destroy or undermine natural ecosystems and that encourage or facilitate long-term growth and development patterns that expose more people and property to hazards.
- Substantial attention must be paid during rebuilding to future potential losses from hazards.
- The principal resources available for recovery linked to sustainability are the people themselves and their local knowledge and expertise. By mobilizing this resource, positive results can be achieved with modest outside assistance.
- Technical assistance and training from outside organizations serve to reinforce and strengthen local organizational capacity, which in turn permits the implementation of recovery, reconstruction, mitigation, and various developmental activities.
- Postdisaster construction and land-use policies must recognize that natural disasters are recurrent events in natural ecological cycles and thus impose limits on redevelopment.
- Features of the natural environment that serve important mitigation functions, such as wetlands and sand dunes, should be taken into account during rebuilding.
- Scientific uncertainty about occurrence, prediction, or vulnerability should not postpone structural strengthening or hazard avoidance during rebuilding.
- Postdisaster reconstruction that does not account for future natural disasters is an inefficient investment of recovery resources.
- Individuals and groups across all sectors should participate in recovery planning.

CONCLUSIONS

Preparedness and Response

Four developments stand out from the research and experience of the past two decades. First, effective preparedness and response activities help save lives, reduce injuries, limit property damage, and minimize all sorts of disruptions that disasters cause. As such, preparedness and response are vital to society's ability to survive extreme natural and technological events over the long term. This contribution to disaster resiliency is by far the strongest link between preparedness and response and sustainable hazards mitigation as described in Chapters 1 and 9.

Second, the theoretical approach to disaster preparedness and response (and research on it) has changed dramatically in the past 20 years. It has moved from a "functional" view of disasters to one that recognizes the tremendous influence social norms and public perceptions and expectations have on the occurrence, effects of, and recovery from disasters. This broader view, which acknowledges the variability and subjectivity of people's outlooks, will be invaluable as the U.S. population becomes more diverse. There are many details to flesh out to complete the understanding of the varied ways in which different people or groups prepare and respond to disasters. In addition, there is much to learn from research done in other countries and from cross-cultural work.

Third, a great deal has been learned about who prepares for disasters (households with higher socioeconomic status and those of nonminorities), but why they do so is still a mystery. Socioeconomic differences are now recognized to play a large role in determining whether and how people get ready for disasters and react once they have occurred. There is considerable understanding of the way information about hazards and preparedness recommendations can foster appropriate behavior. But knowledge about all of these issues and factors is incomplete.

Finally, it is hoped that over the past 20 years the myth of human dysfunction in the immediate emergency period after disasters has been thoroughly dispelled. In case after case people have shown that their response to disasters is characterized by altruism, an enhanced sense of community, helpfulness, resourcefulness, and extraordinary resiliency.

Recovery

There has been a shift in conceptualizing disaster recovery since the first assessment. Recovery was viewed then as a linear phenomenon with stages and end products. These approaches have been augmented by viewing recovery as a process of interaction and decisionmaking among a variety of groups and institutions, including households, organizations, businesses, the broader community, and society. A further shift may be needed—that of giving less attention to restoring damaged structures and more attention to decisionmaking processes at all levels as they relate to the recovery process.

Research has documented that most local disaster plans need to be extended to more directly and explicitly address recovery and reconstruction. There are many postdisaster opportunities for rebuilding in safer ways and in safer places. Disaster plans could start to identify such opportunities in advance.

A growing number of studies have found that locally based recovery approaches are most effective. The definitive characteristic of the bottom-up approaches is that the principal responsibility and authority for carrying out the program rest with a community-based organization, supplemented as needed by technical and financial assistance from outside the community.

CHAPTER EIGHT

Innovative Paths and New Directions

VARIETY OF IDEAS, TECHNOLOGIES, and approaches for hazards and disaster management have emerged since the nation's first assessment (White and Haas, 1975). Many of these have had great impact on the nation's efforts to deal with hazards over the past two decades. Moreover, some of them hold great promise for the development and implementation of sustainable hazards mitigation because they point to ways to obtain better data and information, provide for alternative future loss scenarios through modeling and decision support systems, and allow for greater linkages between research and practice through computer-mediated communication and after-action response.

Striving for sustainable hazards mitigation requires results from interdisciplinary and cutting-edge research and then linking researchers and experts, and what they know, with practitioners. This chapter reviews some innovative paths and new directions and points to the future promise that each of them holds for contributing to sustainable hazards mitigation.

NEW TECHNOLOGY AND APPROACHES

Computer-Mediated Communications

The nation's hazards information base has grown enormously since the first assessment. Today there are many more hazard libraries and periodicals, and the number of books published annually on the topic has grown exponentially. But two information innovations have brought about unprecedented changes: computer-mediated communications via the Internet and the World Wide Web (WWW) and the unprecedented amount of information that these modes of delivery and communication make available. As an example, in 1996 the Federal Emergency Management Agency (FEMA) offered more than 8,000 pages of information via the WWW, and the FEMA website was accessed over 5 million times in the previous calendar year; the Natural Hazards Research and Applications Information Center's (NHRAIC) list of useful emergency management websites numbers 80 and dozens of other Internet resources; and the worldwide distribution of NHRAIC's electronic newsletter, "Disaster Research," has grown at a rate of 60 percent per year since it was begun only a few years ago.

Today's computer networks enable researchers, information centers, and any interested persons to rapidly obtain and exchange information. The Internet has provided a means through which people can become participants in a national and world colloquy on hazards and disasters. But beyond this, computer networks help to resolve issues and problems. For example, they consolidate and make more available the information maintained by hazard centers. Internet technology is creating a virtual consolidated library and database of hazards information that includes the collective holdings of many institutions. Additionally, computers and networks have the prospect of making it more feasible to collect, consolidate, and disseminate information germane to sustainable hazards mitigation, such as data on disaster losses and costs called for in Chapter 3 and information relative to determining local vulnerability called for in Chapter 9. And networks are a low-cost way to transmit information about hazards and disasters from the United States to developing nations.

Geographic Information Systems

Geographic information systems (GISs) are computer-based tools and procedures that capture, store, analyze, and display spatially referenced data. Broadly speaking, a GIS is a set of integrated digital maps and spatial information that are electronically linked. GISs have been used for hazards management since the Canadian Geographical Information System was developed in the 1960s for the management of forest inventory and threats to it. It was recognized early on that hazards with a distinct spatial extent, such as floods, were perfect candidates for GISs. Today, almost every imaginable hazard has been examined, mapped, analyzed, modeled, and predicted in GISs by including data layers such as physical hazards processes, impact zones, evacuation routes, damage, and current and proposed mitigation measures. Although GIS analyses to date require refinement and additional data and information input to sharpen their results, GISs hold great potential to inform actions for sustainable hazards mitigation because they can assemble and integrate all of the factors needed to consider alternative short- and long-term hazards and losses.

Applications

GISs have applicability in almost every aspect of hazards and risk management. For example, they can help in preparing evacuation routes, emergency shelter placement, and assignment of given population sizes to shelter capacities. GISs can be used to identify special-needs populations and pickup routes for evacuation. A GIS also fits well into a community's postevent emergency response; for example, the first reports of closed roads and infrastructure damage can be coded into a GIS and rough maps can be made to aid emergency personnel. Such maps can be used to navigate a damaged landscape, which might lack familiar landmarks. The rerouting of access corridors around inaccessible roads to enter the damaged area can be outlined. Damage estimates can be encoded as they are reported. A flyover by aircraft and satellites can add imagery of damaged areas to those estimates. Potentially hazardous structures can be marked and coded.

A GIS can also inform recovery. Damaged areas can be digitized and used to set priorities for such tasks as rebuilding and debris removal. Citizens in neighborhoods that were hard hit can be targeted for counseling, education, and disaster assistance. Areas for potential new business development can be outlined and support maps and statistics created for redevelopment efforts. Finally, a GIS can be especially helpful in informing sustainable hazards mitigation and designing disaster-resilient communities because it can help in the identification of hazards in communities and at project sites, evaluate vulnerability to hazard impacts, and help apply hazard mitigation design solutions.

A Notable Example

Use of a GIS for emergency management after the 1994 Northridge earthquake illustrates the technology's contribution to emergency response. GIS techniques were used by private firms and local, state, and federal government agencies to identify damage and coordinate recovery efforts. Using magnitude and epicenter information, EQE International, on behalf of the state of California, analyzed building inventories, regional population, and economic data for the affected area. This enabled an intensity map to be generated, which helped target emergency response to the areas of heaviest damage. Property loss estimates were incorporated, enabling a total loss estimation to be forwarded to FEMA. The GIS was also used to determine optimal sites for disaster assistance field offices, to track the repair of damaged buildings, and to determine cost-effective mitigation measures for schools, hospitals, and other public facilities. Although it was not without problems, a GIS became part of the near-real-time emergency response after the Northridge earthquake.

Future Possibilities

The future will expand the use of GISs in hazards and risk management. Systems specifically tailored for emergency managers are being marketed, while the entire GIS industry is becoming more entrenched. Several trends point to this strengthening, including analysis of real-time data, verification of hazards and vulnerability models, and more spatial analysis tools allowing emergency managers to keep up with rapidly changing situations and immediately incorporate them into response plans. Direct connections between handheld positioning units and the central GIS at command-and-control centers will allow managers to direct rescue and cleanup operations with more accuracy and precision.

As more potential damage models are developed and community vulnerability maps are created, the opportunity to test these predictions becomes more likely and GISs will be of use in determining the strengths and weaknesses of damage predictions. Finally, the increasingly sophisticated modeling ability of GISs will allow more hazard and risk managers to ask "what if" questions about future losses in both the short and the long term in order to guide today's mitigation decisionmaking. A GIS

This highway near Los Angeles was damaged in the Northridge earthquake of January 17, 1994. Photograph by Ted Soqui, copyright Sygma.

enables the consolidation of data and information from the full range of disciplines and areas (e.g., natural science, social science, engineering) to consider future impacts and loss.

Remote Sensing

Enormous growth in satellite remote sensing capability has developed since completion of the first assessment, and there has been an increase in the use of remote sensing for disaster management. The future holds even greater promise, particularly for satellite remote sensing for sustainable hazards mitigation. Broadly defined, remote sensing includes sensors placed on fixed-wing aircraft and satellites. Fixed-wing platforms can be positioned on demand and are a good source of postimpact data relevant to disaster management.

Until recently, there were limitations to the use of satellite remote sensing by disaster managers. Most high-resolution earth observing systems have been experimental, and many organizations and agencies have been reluctant to invest the startup time and funding required. This problem may lessen with the projected growth of the satellite remote sensing industry, which will offer ultra-high resolution imagery on demand. There is also an emerging support industry for interpreting imagery and concatenating it with other data through a GIS. Finally, classified remote sensing systems can be used to create products for postdisaster assessments, but these data are not yet generally available. Many vulnerability and damage assessment studies will continue to rely on airborne systems, which currently offer the only tool that may be positioned immediately after an event or at an altitude for sufficient spatial resolution.

The increased use and utility of GISs and their applicability to sustainable hazards mitigation, coupled with the availability of relevant databases, also support the likely future growth of remote sensing applications. GISs provide a receptive format demanding up-to-date information that often can be provided by remotely sensed larger-scale hazards-related information about natural systems. Interpreting remotely sensed data is more complicated than a simple photo interpretation. For example, the accuracy of a floodplain map depends on more than remotely sensed imagery, and an accurate model would also include GIS-based data on runoff, soils and percolation, precipitation inputs, and other information.

Current Capabilities

Optical instruments are flown in satellites in a polar orbit that crosses the equator at the same time of day so that they cover the entire earth approximately every two weeks. Cloud cover cuts back the observational frequency to four weeks on average, but some instruments are able to view the side of the satellite track, thus improving the frequency to three or four days. Spatial resolution is currently as low as 5 meters for the Russian satellites and is due to decrease to 1 meter with new U.S. commercial satellites. U.S. polar-orbiting meteorological satellites carry an instrument that can make daily earth observations (at low spatial resolution, 1 to 4 kilometers) and are used to monitor vegetation and volcanic clouds.

Optical geosynchronous (meteorological) satellites hover over an equatorial spot on earth, permitting continuous hourly (albeit low-resolution, 4-kilometer) observations. These data now are provided by vendors in the United States, Japan, and Europe.

Synthetic aperture radar can penetrate cloud cover, affording more frequent coverage. In addition, radar can improve on optical observations of soil moisture, for example. Systems from Europe, Canada, and

Japan are currently in operation. The Canadian system is commercial and has a capability of side-to-side tracking, thus improving the frequency of observations to several days. Passive microwave sensors can measure and map the natural microwaves emitted because of the earth's temperature. The U.S. Department of Defense carries a passive system that, among other things, can be used for estimating soil moisture.

Historical Applications

During the past 30 years, remote sensing has been used to monitor and map hurricanes, lava flows, landslides, wildfires, and biomass burning. The first satellite monitoring of a hurricane was Hurricane Camille in 1969. Remote sensing has also been used to inventory hazardous waste disposal sites and air pollution, including radioactive releases.

Postevent assessments using remote sensing have been used in oil spills, hurricane-induced damage, and floods and were especially useful in gauging the extent of the 1993 Mississippi River floods and the spatial and financial extent of damage from Hurricane Andrew.

Land-use and land-use change maps are fundamental elements of vulnerability assessments. In areas of rapid population growth, such as coastal counties, updating these maps is facilitated by newly available commercial high-resolution remotely sensed data. Data at lower resolutions are used to update land cover and topographic maps, especially for surface water runoff. Technology now exists to provide such data at an accuracy of less than 1 vertical meter.

Most applications of remote sensing for disaster preparedness require appropriate data and effective models for generating forecasts and warnings. The lack of good models of biophysical, geophysical, and social systems models is probably the biggest impediment to the wider application of remote sensing to disaster management. Although hurricane and flood forecasts are fairly successful, there are still errors largely because of the accuracy and reliability of the models. Improved modeling and understanding of the El Niño phenomenon, based on remote sensing data, is extending long-term warning of droughts, floods, and severe storm frequency. Finally, remote sensing data can provide important input into damage assessment models.

With a few exceptions, the immediate response phase of a disaster places the most severe demand on remote sensing systems in terms of temporal and spatial resolution. Ideally, emergency managers need specific, detailed, and timely damage reports, which generally cannot be provided by satellites except under extremely fortuitous conditions. Instead, aircraft observations are used. Both film and digital sensors are used on fixed-winged aircraft and linked to global positioning system locators. Digital maps of the phenomenon may be created by conversion of the film to digital format. In this way these georeferenced high-resolution digital maps can be incorporated into a GIS. Air-to-ground communications systems have been developed that permit real-time transmission and application of these data. In the future it might be possible to use such observation/communications systems deployed on remotely piloted vehicles that can be launched and controlled directly from a disaster site.

Decision Support Systems

Decision support systems (DSSs) are computer information systems that have evolved from existing but disparate information pathways used by decisionmakers. Technical and other support for decisions has always existed in familiar forms—for example, reports, maps, raw data, and statistics. Computer technology now makes it possible to create, analyze, and store these different forms of information in an integrated system with four components: control system, database system, model, and report system. The computer allows for instantaneous links among the database, the GIS, simulation models, and management tools. Other components may be part of a DSS, such as optimization routines for resource allocation or artificial intelligence and expert systems.

DSSs have great potential for facilitating sustainable hazards mitigation because they can be linked with simulation models (or economic effects, hypothetical physical phenomena, social response), and both can be tied to a core database housing information on infrastructure, demographics, risk, vulnerability, and potential response alternatives. Ultimately, the DSS ties together spatial and temporally based information with simulation models that depict a spatial/temporal reaction to a response and allows problem-specific information to be reported. For example, expected losses could be estimated in alternative local hazard scenarios in the short and long terms. In the process, local decisionmakers could "see" the future losses and impacts of the decisions they make today. This would help them design futures within locally determined bounds of acceptable losses.

The ability of DSSs to tie together information common to long- and short-term planning fills a gap in hazards management. A common tool that would house historical data (losses, physical events, trends), current data (weather, demographics, political information), simulation tools (environmental, economic, and socio-political models), and report tools (statistical analyses, maps) promises to improve hazards-related decision-making.

Risk Analysis

Modern risk analysis has its roots in probability theory and in the identification of causal linkages between adverse health effects and hazardous activities in the 1940s. Risk analysis encompasses risk assessment (where there is uncertainty by scientists) and risk management (which is more policy driven in the evaluation of alternative regulatory strategies). Recently, risk management has been broadened to include a range of scientifically sound cost-effective actions that take into account ethical, socio-cultural, political, and legal considerations (U.S. Presidential and Congressional Commission on Risk Assessment and Risk Management, 1997).

Starr's (1969) work on social benefits versus technological risk began quantitative risk analysis and the development of the risk analysis paradigm. The paradigm gained wide acceptance in 1983 when the National Research Council recommended a risk assessment framework for the federal government (National Research Council 1983). A short while later the U.S. Environmental Protection Agency (1993) developed a comparative risk analysis approach. In 1994 Congress established the Commission on Risk Analysis and Risk Management and asked it to modify federal approaches used to assess and manage risks. A revised framework was issued in 1997 that includes: (1) problem definition and context, (2) analysis of associated risks, (3) examination of options for addressing the risks, (4) decisionmaking about which options to implement, (5) implementation of decisions, and (6) evaluation of the actions taken and their results. The framework is conducted in collaboration with stakeholders and includes feedback that permits incorporating new information into the framework.

Uncertainty and Determining Acceptable Risk

The two challenges in risk analysis are how to handle uncertainty in knowledge about risk and determining how safe is safe enough (Kunreuther and Slovic, 1996). The diversity of methods used in risk analysis prompts many to describe it as more of an art than a science. The lack of conformity of data for comparing risks generates some of the greatest concerns. Often assessments give the impression that there is more knowledge and certainty than actually exists. On the other hand, they also may suggest that uncertainty is a function of measurement, when in fact it may be a difference in professional judgment, not in the lack of data or the quality of the data. Uncertainty also gives the impression that certain risks exist when they often have not even been analyzed. Scientific uncertainty then creates a milieu in which risks can either be overstated or understated depending on motives or choice (NRC, 1994).

The second issue is what constitutes acceptable risk. Clearly, this is a more policy-driven issue than a scientific one; however, science is often used to justify the decision. The use of cost-benefit analyses (now renamed risk-benefit) is one example. In engineering a popular way to define acceptable risk is to specify a recurrence interval, which expresses the failure rate of the systems. Failure rates are often determined by explicit or implicit cost-benefit risk analysis. For example, the appropriate level of protection for agricultural levees is determined by maximizing net benefits and considering incremental crop damage versus the added cost of raising a levee. Another method is to utilize low-probability/high-consequence versus high-probability/low-consequence determinations and plan accordingly.

Stakeholder Involvement

Involving affected populations has always been implied but has not been explicitly mandated until recently. The rapid evolution of risk communication as a field of study has helped incorporate stakeholder involvement in risk management. The number and types of stakeholders depend on the situation but typically include those groups that are affected or potentially affected by any effort to manage the risk in question. Stakeholder involvement should be a part of not only the action but, more importantly, the identification and assessment phases. In this way, differing technical assessments can be considered along with public values, perceptions, and knowledge. It takes more time and money to involve stakeholders in the risk analysis process, but the long-term savings more than compensate, especially in the acceptability of resulting mitigation options.

Applications to Sustainable Hazards Mitigation

Risk analysis could be used in concert with other techniques such as GISs, DSSs, and remote sensing to generate alternative short- and long-term development and hazard scenarios for the nation's communities and projects within them. When using risk analysis, the scientific and political bases for the definition, analysis, and options should be made explicit to all stakeholders. Sensitivity analyses should be done to show where additional research might resolve some of the uncertainties and where it would not help at all. Decisionmakers should be provided with both quantitative and descriptive risk characterizations to enhance their understanding of an issue and how it fits into the larger context. Movement away from a single-risk or single-hazard focus to a multirisk or multihazards approach is needed to foster decisionmakers' understanding and informed choices about sustainable hazards mitigation.

Loss Estimate Modeling

Computer aided modeling software to estimate losses in future disastrous events has emerged since the first assessment. Its development and level of sophistication continues to grow rapidly, but its prime promise for sustainable hazards mitigation lies in developing its ability to provide local decisionmakers with specific information to inform local decisions. Much more work on loss estimate models is needed to increase their local utility. A variety of loss estimate models currently exist. The three described below illustrate the promise of such techniques.

HAZUS

HAZUS operates through MapInfo—a GIS application—to estimate losses in future earthquakes, although work has begun to extend its application to other hazards. It is being developed by the National Institute of Building Sciences with funding from the Federal Emergency Management Agency (FEMA). HAZUS uses mathematical formulas and information about building stock, economic data, local geology, the location and size of potential earthquakes, and other information to estimate potential losses. HAZUS is able to map and display ground shaking, patterns of building damage, and demographic information about specific communities. Loss estimates include the number of buildings damaged, the amount of damage to transportation systems and electrical and water utilities, the number of casualties, the number of people displaced from their homes, and the estimated costs of repairing the damages.

HAZUS has applications both before and after an earthquake. Prior to an earthquake, the loss estimates can assist with mitigation policy making, such as helping with land use decisions, zoning, and building codes. It can also be used for developing emergency response plans and for testing response capability. HAZUS can also assist in decisionmaking immediately following an earthquake, as it can identify the most likely hit areas, provide a fast estimate of the damage and casualties, and determine the resources that need to be sent to the affected area. During the response period HAZUS can help project the number of homeless, the types of financial and material resources victims will need, and estimate medical demands.

TAOS

TAOS, which stands for "The Arbiter of Storms," can examine and simultaneously integrate the effects of multiple natural hazards such as wind, waves, storm surge, and inland flooding from rainfall. TAOS takes into account the category of the storm; its forward movement, wind speed, maximum radius of winds, and direction; land cover characteristics; and the effects of urbanization. It is currently used in Florida to produce a consistent statewide assessment of storm hazard risk. TAOS's results serve as the basis for developing local mitigation strategies, and provide dollar estimates of potential property losses throughout the state. TAOS uses information such as historical and hypothetical storm parameters, land cover information, and model forecast data from the National Weather Service National Hurricane Center. Since the TAOS model relies primarily on public domain information from standardized databases, it can be easily updated. TAOS also is compatible with most Geographic Information System (GIS) programs. The model's current utility for mitigation rests on its ability to provide maximum water level, maximum wind speeds, maximum inland flood water height from rainfall, maximum wave heights, structural damage estimates, and debris estimates.

IRAS

The Insurance and Investment Risk Assessment System, or IRAS, is a model used to analyze financial risk due to natural hazards and is used by financial institutions, such as insurance firms, banks, insurance brokers, mortgage institutions, and government agencies. IRAS is used for risk engineering and mitigation, insurance underwriting, and for risk selection. IRAS uses proprietary databases to model financial risk, and is divided into three subsystems—hazard, vulnerability, and financial. The hazard and vulnerability modeling are peril and geography specific. A process called geocoding provides the latitude and longitude of specific addresses and is used to gather additional information about the site. During a hazard analysis, the model simulates specific events and determines the likely impact at each site. After the impact has been calculated, the vulnerability model estimates physical damage, by using data such as occupancy type, construction class, age of construction, and number of stories. Finally, after the physical damage has been calculated, the financial model determines losses to the participants in the event by calculating complex financial structures. IRAS does this by taking into account factors such as the financial terms and conditions for each individual policy, including coverage limits and deductibles, and all per risk reinsurance certificates. The financial model is able to analyze losses from the perspective of any insurance or reinsurance participant.

Postdisaster Research

Several organizations and agencies have programs to gather knowledge immediately after a disaster. For example, the Earthquake Engineering Research Institute (EERI) conducts a "Learning from Earthquakes" program in which teams of engineers, physical scientists, and social scientists enter the field immediately after an earthquake to learn perishable lessons. The Natural Hazards Research and Applications Information Center sponsors a "Quick Response" program to enable social scientists to enter the field after disasters to collect perishable data on issues ranging from organizational and individual responses to a disaster to the effectiveness of warning systems to the economic impact of a particular event. Both programs produce written reports of findings and recommendations for changes.

Federal agencies also undertake such efforts. The U.S. Army Corps of Engineers often conducts a reconnaissance study after a disaster and writes an "After Action" report to assess the effectiveness of its response, identify any problems, and recommend improvements. Similarly, the National Oceanic and Atmospheric Administration takes a team of physical and social scientists into the field after a major severe weather event to assess how well the event was predicted and forecast and how well the warnings were issued, disseminated, and received. Knowledge gained from such efforts has been useful in adjusting policies and day-to-day practices for National Weather Service personnel. Under the National Oil and Hazardous Substances Pollution Contingency Plan, the U.S. Environmental Protection Agency has organized the National Response Team (NRT), which investigates the release of hazardous materials that pose environmental threats and result in significant natural resource damage and makes recommendations for improving interagency coordination. The NRT and its member agencies have extensive experience in preparedness and response for oil and hazardous materials spills.

Perhaps the most ambitious effort along these lines has been institutionalization of the Interagency Hazard Mitigation Teams (IHMTs) organized by FEMA after major disasters. Originated in the early 1980s and initially used only after floods, the IHMTs now gather representatives of all federal, state, and local governments involved in a particular disaster to identify opportunities for mitigation during the recovery period. The team prepares a report 15 days later, listing possible mitigation projects, funding sources for them, and responsible parties for follow through on each. The idea was to put together the list of recommendations quickly so that they could be incorporated into the reconstruction process. The team also must prepare 45- and 90-day reports to track progress on implementation of the mitigation recommendations. Over the past 17 years, the use of IHMTs has expanded to all types of disasters. Dozens if not hundreds of 15-day reports have been written listing thousands of recommendations for improved mitigation in communities across the nation. The effectiveness of this approach is difficult to judge. There is anecdotal evidence that report recommendations were sometimes followed, but no systematic analysis has been performed.

The postdisaster research discussed above is likely the least expensive and most effective hazards and disaster research conducted in the nation, regardless of program or sponsoring organization. So it is difficult to speculate on how to improve its cost effectiveness. But it is likely that those who engage in postdisaster research and "After Action" reports could benefit from a workshop to examine how their research agendas, team participants, and research styles may or may not change under a sustainable hazards mitigation approach.

THE RESEARCH INFRASTRUCTURE

Future Research

Sustainable hazards mitigation will demand that future researchers work in a world different from that of the past and different even from today. The questions they address and the solutions they seek will be more complex; they will work less in disciplinary isolation and more in interdisciplinary teams; and they will work with practitioners more frequently and intensively than is the case today. Moreover, there may well be fewer dollars to support their work.

Funding

Government support for research is under increased scrutiny. It was primarily through government grants that past researchers received financial support for knowledge generation. Early support for hazards and disaster research produced today's research centers, and government support has been the financial locomotive for training young researchers, maintaining the nation's research infrastructure, and producing new knowledge. Today, the private sector is being asked to fill the gap between government support for research and development and projected research and knowledge needs.

Complexity

Future researchers will be confronted by increased complexity in the subjects they investigate. Simple linear models of environmental cause and effect will no longer be adequate. Higher densities of at-risk populations will increase the stakes for useful findings, and national and global calls for sustainability will grow louder. Hazards researchers are being called on to make sense out of, and offer solutions to, complex macro-level social problems. And they will be confronted with investigating "chain reaction" disasters as the growing size of catastrophic events no longer affects just municipalities but impacts megacities, regions, and entire sectors of the nation.

Technological advances in the nation's constructed environment will continue to escalate as improvements in building codes and materials result in more advanced structures. This too will add to the complexity of future research because contemporary technological developments require more resources to keep systems functioning. System sophistication results in increased vulnerability, and entropy research will have to be addressed in the context of already complicated and difficult to measure cost-benefit analyses. It will be increasingly difficult to have an overview of all of the parts of a problem in a complex society.

A Problem Focus

Society will demand more effective use of public funds, and this will require researchers to produce even more applied and directly useful findings. This demand is not new: hazards and disaster research over the past two decades is a model of how scientific and engineering research can be organized to help solve social problems. But the demand will grow louder. The research community will no doubt respond, but it is unclear how the increased demand will impact the nation's research universities, which direct their faculty—at least those who would keep their jobs—away from problem solving and into the theoretical corners of distinct disciplines.

Several outcomes are possible. Hazards research could shift away from universities into the hands of private-sector entrepreneurs, or the reward system in universities might shift to accommodate the demand for more useful research results. There is evidence that both shifts are occurring.

The University Reward System

Faculty in the nation's research universities are evaluated, retained, and rewarded on the basis of publishing new knowledge that furthers the basic theories of their disciplines, not for the application of that knowledge to solve society's problems, for the effectiveness of their teaching, or for the service they render to on- and off-campus communities. What distinguishes institutions of higher learning from each other is not which of these criteria is used to evaluate faculty but rather which is emphasized. The mix used in any one campus directs the faculty members' focus and creates a national hierarchy of prestige among the nation's colleges and universities.

Faculty-generated grants and contracts enhance individual faculty prestige, provide additional funding for the university, and help attract and support quality graduate students. At large research universities, successful grant writing is not optional. A portion of all grant monies (perhaps 45 percent of every dollar and called "overhead") is subtracted from grants and contracts that are awarded; it is kept by the host institution before the research dollars are passed on to the researcher. This translates into an indirect payment to the university's operating budget. It is compensation for the space used for research, the resources that faculty members use in securing the grant, and so on.

State legislatures are scrutinizing higher education's priorities as part of their quest for leaner state budgets. Smaller budgets bring new pressures to the university reward system. More faculty members are competing for a smaller number of smaller grants. Heavier teaching loads and less time for research are becoming more common. This has been met with faculty resistance, but budget realities make it an attractive option to legislatures and institutions in most states. The number of tenure-track faculty jobs is on the decline, replaced by temporary teaching-only faculty.

Herein lies the current dilemma for the future of hazards research in U.S. universities: the nation and its research funding agencies are demanding more practical applied research, but the nation's young research faculty members are under ever-increasing pressure to conduct more theoretical and nonapplied work if they are to keep their jobs. This conflict has always been present, but it has never been as intense as it is today. The resulting tension has contributed to the recent tendency of young hazards-trained faculty members to leave the hazards research area in favor of activities more closely linked to job security.

There is a mismatch between the way universities are currently organized around traditional disciplines and the interdisciplinary problems that society now faces. Perhaps there will soon be a birth of new university-based programs and disciplines to respond to the needs of society. For example, interdisciplinary environmental studies programs are being invented on many campuses, and emergency management, hazards, and disaster degree programs are coming into existence. New university-based disciplines to address sustainable development and sustainable hazards mitigation may be invented, and it may well be those disciplines that will house the nation's next generation of hazards and disasters researchers.

National Social Science Data Repository

There is no central repository for the other social response data collected in the nation (although a few institutions maintain limited collections of their own materials). Researchers should be encouraged to archive datasets for others and for the long term in a central repository within a year or two after collection. Although some funding groups, such as the National Science Foundation (NSF), have a policy that encourages researchers to put data into the public domain within two years of its collection, it is not clear that this policy is enforced. Researchers should be asked to file copies of data collection instruments, specifications regarding how and why the data were collected, and copies of proposals with the original data into a central repository so that the data are permanently available to other researchers and future generations.

A Centralized Archive

Some ongoing funding would be needed to establish and maintain a central location for maintaining archived data, but the information itself might be made available through a website or other computer-accessible resource. A standardized mechanism would have to be developed to ensure that those who use the data credit its originator. Computer records should be kept on each person or group that requests information about or uses data from the archive. Such an archive could be expanded over time to include all related datasets and could even be a resource for practitioners and policymakers. It also might be used to encourage government agencies to centralize data collection.

Standardized Measures

The existence of a central archive would accelerate the development of standardized measures of data collection on the social aspects of hazards and disasters. Currently, across-disaster comparisons are constrained because different researchers measure the same things in different ways. No central location exists from which a researcher can find and obtain measures for a particular aspect of a disaster—for example, applications to agencies for assistance after a disaster, how well existing measures worked in getting needed information, or the extent to which the research group that designed it would recommend that it be adapted. Sustainable hazards mitigation would be facilitated by standardized measurement in order to provide for across-community comparisons.

EDUCATION

A variety of initiatives are either under way or have recently been proposed to expand education for hazards mitigation and emergency preparedness. For example, the first text on emergency management was published by the International City/County Management Association in 1991 (see Drabek and Hoetmer, 1991). Sustainable hazards mitigation will require that some future hazards managers be trained across traditional disciplines and in interdisciplinary ways that would enable them to view the world that they manage holistically. Some initiatives in place foretell this alternative type of education, and other approaches must still be invented.

Higher Education Initiatives in Place

FEMA's Emergency Management Institute (EMI) has undertaken several recent projects to promote college-based emergency management education. Most unique is EMI's current initiative to work with academics to develop up to two dozen courses related to hazards for the college and university levels. Each will be in the form of an instructor guide in sufficient detail to enable an informed professor to teach the course with a minimum of additional work. Professors will be provided with a disk copy and encouraged to make appropriate modifications. There will be no reporting requirements and the materials will be free. FEMA wants to see an emergency management degree program in every state by the year 2000, including associate degrees, baccalaureate programs, and eventually graduate degrees.

Higher Education Initiatives Needed

Two new initiatives are needed to support sustainable hazards mitigation. First, the holistic perspectives needed for sustainable hazards mitigation cannot be gained through education in just any one of the specialized fields available today. Some members of the next generation must be educated across the disciplines applicable to sustainable hazards mitigation—for example, different fields of engineering, physical sciences, and social sciences. Interdisciplinary problem-focused degree programs would provide professionals with the tools needed to access new knowledge from those educated in more traditional ways and would facilitate the application of interdisciplinary solutions to tomorrow's problems. If they had their own faculty and reward systems, these interdisciplinary degree programs would do much to alleviate the current pressure on faculty to stay clear of applied work.

Second, social science researchers engaged in hazards and disasters studies should collaborate to create a university without walls. This network of colleagues would facilitate cooperative projects; enhance the sophistication and breadth of the research produced; encourage standardization of instrumentation, data processing, data documentation, and analytical approaches; and increase interaction. But existing barriers discourage such efforts. For example, the academic community recognizes individuals but is hard pressed to do the same for interdisciplinary groups; promotion committees have difficulty ascertaining the relative contributions in multiauthored publications; the overhead structures of many institutions discourage cross-institutional research teams; and graduate students are restricted to the department and university in which they are enrolled.

Innovative approaches are needed to overcome these and other barriers to bring researchers together. A recent example was an NSF-sponsored project that brought faculty members and students from different universities and disciplines together to work on a common problem. A university without walls would position social science researchers to contribute more fully to the blending of traditional disaster research with broader related areas—for example, environmental science and management, planning, and technology—in ways that will foster sustainable hazards mitigation.

Outreach to Professional Groups

Technology transfer moves knowledge and know-how from one or more professionals to other professionals to achieve practical purposes: reduction of risk from hazards or increased preparedness for disaster response and recovery. It is difficult to achieve because it requires interaction between "internal" and "external" champions and calls for new ways of thinking and acting. Currently, technology is transferred more effectively in a crisis environment when the need for the knowledge is obviously high—for example, after a natural disaster or following a new political initiative—than in a "normal" environment. This will have to be changed for sustainable hazards mitigation to become a reality.

The process for transfer of technology to professional groups has been examined extensively both in the United States and abroad since 1977 through the National Earthquake Program (see Hays and Rouhban, 1991). For example, a series of 45 workshops convened in every earthquake-prone area of the nation brought together hundreds of researchers and practitioners annually for two to five years and established a process for fostering more knowledgeable use of science and technology in policy decisions. The process was realized most rapidly in California, and few policies were changed outside California unless an earthquake disaster occurred. The same experience with technology transfer has been repeated elsewhere and by others.

There is much to be gained from further research on technology transfer, which requires a sustained long-term effort built around opportunities. For example, the results of postearthquake investigations of the 1995 earthquake in Kobe, Japan, are just now emerging. Japan and the United States are developing new mechanisms and improving existing ones for cooperating and transferring technology. The effects of this effort will bring new insights on technology transfer in two distinctly different cultures. The continuation of postdisaster investigations will always provide new opportunities to accelerate technology transfer.

INTERNATIONAL COMPARISONS AND EXCHANGES

Although comparative international research is relatively scant, some agencies, some hazards (especially earthquakes), and some issues (e.g., vulnerability, warning, relief, information systems) have received growing international attention since the nation's first assessment on natural hazards (White and Haas, 1975).

International Comparisons

Despite the impact of globalization and global change on the hazardousness of the United States, the nation's agencies have historically been limited to concentrating on hazards only in this country. Although U.S. scientists do communicate with their counterparts in other countries and participate in collaborative projects, the limited perspective of U.S. agencies continues to be evident in the publications and new electronic libraries of FEMA and other agencies. (Some exceptions are the recent U.S.-Japan cooperation on the Natural Disaster Reference Database and National Research Institute for Earth Sciences and Disaster Prevention.) Minimal funding for the Office of Foreign Disaster Assistance of the State Department has limited that organization to occasional relief activities with no budget for research. Although some research is done by in the U.S. Agency for International Development, it has not been a focused priority. U.S. scientists and scientific organizations have a stronger international perspective, but the circulation and application of that knowledge in the nation remain constrained by the domestic scope of federal agency programs.

Some 30 years ago a consortia of hazards researchers began a program of comparative international studies. Major disasters, such as the Bengal cyclone of 1970 and the famine in the Sahel in the early 1970s, raised international concern to new levels. The results of those early efforts are summarized in two volumes, one by White (1974) and the other by Burton et al. (1978). About the consortia's purpose, White (1974, p. 13) wrote:

> The first lesson is that if costly threats to life and property from the extremes of natural phenomena are to be minimized, there must be careful sharing of the skills, experience, and research capacity of the family of nations.

Comparative international research has been performed since those words were written. Perhaps the most serious social science effort was made by the Disaster Research Center at the University of Delaware. The Earthquake Engineering Research Institute has conducted systematic international studies of earthquakes wherever they happen around the world, as has the U.S. Geological Survey. Nevertheless, international research will need to play a larger role in the future if the United States is to capture the implications of global systems for sustainable hazards mitigation.

A key research issue for the beginning of the twenty-first century is how U.S. agencies, scientists, and managers can draw effectively on the escalating wealth of "skills, experience, and research capacity of the family of nations" mentioned by White. The past 20 years of research have identified shared concerns, academic pitfalls, and promising approaches and methods. The logic, methods, and policy applications of comparative international research remain, however, at a rudimentary level of development.

International Exchanges

The United States should share knowledge and technology related to sustainable hazards mitigation with other nations. Four points seem obvious. First, disaster aid should not come in the form of increased debt. The prime motive for sharing sustainable hazards mitigation information and technology should be improving quality of life for the largest number of people in the most equitable way since these are two primary elements of sustainable hazards mitigation. Help with economic growth and disaster resiliency should consider the manner in which the benefits of growth are to be redistributed and how they will contribute to quality of life in the long run.

Second, disaster experts and development experts should work together to address the root causes of disaster vulnerability, including overgrazing, deforestation, poor irrigation management, intensive farming, overdevelopment, unplanned development, environmental degradation, increased industrial activity, the intensive (and often inappropriate) use of modern technologies, making plans based on political boundaries rather than ecological or geological boundaries, poverty, poorly functioning public services, unwise political decisions, and lack of contact between people working on development and those working on disaster problems.

Third, development and disaster experts from developed countries have a lot to learn from previously ignored populations in developing countries. For example, the western practice of increasing overconsumption, often with no real quality of life improvements, may not be a good model for other countries. In fact, the West should learn from countries that have managed to improve quality of life without significantly increasing their consumption of natural resources. In addition, local traditions and knowledge of past disasters should be seen as useful input into the planning or relief process. Giving women a stronger voice in development and disaster mitigation decisions could improve projects, and make women and children less vulnerable than they are now. Knowledge from developed countries should, of course, be shared with developing countries. However, knowledge should not be forced but shared, with the understanding that it may need to be modified to fit local needs. There is a fine line between sharing new information and respecting local traditions and understandings.

Fourth, there needs to be a general move toward integration of work being done. Economic development projects should consider disaster vulnerability, and disaster projects should consider development patterns. Disaster reduction should be a part of everyday development processes. There needs to be more collaboration between those working on housing, disaster, poverty, education, gender, human rights, ecology, insurance, city planning, and land development. There is a need for institutionalizing these links among and between governmental sectors, nongovernmental organizations, activists, volunteers, scientists, and the private sector.

MODEL COMMUNITIES

One way to foster a shift in the approach that the United States uses to manage natural and technological hazards is to begin the long-term process of creating example communities to show what a locality might look like if it redesigned itself in terms of sustainable hazards mitigation. Others have called for a national investment in prototypes to try out these new ideas.

The Institute for Business and Home Safety's (IBHS) Community Showcase Program is a national initiative to support the development of communities to demonstrate adoption of the latest building codes without modification. IBHS, with its partners, provides direct support to enable a showcase community to implement the provisions of its mitigation strategy. Participation in the showcase community program allows a community to take advantage of incentives and cost savings—for example, possible insurance premium reductions for specific mitigation actions, low-interest loans for retrofitting, and others. The initiative emphasizes structural approaches to mitigation. The Wharton School's Model Cities Program offers a broader approach and includes more than structural approaches.

"Disaster-resistant communities," a term first used by Donald Geis of the International City/County Management Association, has recently been adopted by other groups, such as the Central U.S. Earthquake Consortium (CUSEC) and FEMA. Disaster-resistant communities are defined by CUSEC and FEMA in much the same way that disaster-resilient communities are defined in this volume.[1] CUSEC's call for such communities is based its own call for a paradigmatic national natural hazards shift:

> In essence, we have the knowledge and tools to reduce the vulnerability of our communities to natural hazards. What is needed is a fundamental shift in public perception of natural hazards. Hazard reduction policies and practices need to be integrated into the mainstream of community and business activities throughout the . . . United States. . . . Furthermore, these mitigation policies and programs should be compatible with community goals, as reflected in local comprehensive plans (CUSEC, 1997, p. 5.)

[1]Throughout this book the term "resiliency" to disasters is used rather than "resistance" to disasters because of a sense that resiliency has a slightly broader, more flexible connotation. However, in the context of an entity's readiness and ability to withstand and recover from a hazard or disaster without serious impacts, the two terms can be considered essentially interchangeable.

FEMA has made the creation of prototype disaster-resistant communities the focus of its Project Impact. This is the nation's most serious current attempt to start the conversation for sustainable hazards mitigation at selected locations throughout the country. First, FEMA established a predisaster mitigation fund to provide financial incentives for communities to reduce the vulnerability of their buildings and infrastructure. The agency also is promoting public-private partnerships to develop closer working relationships with the business community in carrying out mitigation efforts. Finally, FEMA is streamlining its public assistance programs to reduce recovery time.

CONCLUSION

There have been many changes in hazards-relevant technologies and approaches since the first assessment was completed. For example, computer-mediated communications, geographic information systems, remote sensing, decision support systems, and risk analysis all have been substantially developed in the past two decades. Each of these shows great promise for contributing to sustainable hazards mitigation.

The research infrastructure has also changed substantially, but the biggest changes in approach and capacity for hazards research have only just begun. Examples include funding shifts, increased complexity in the problems that researchers investigate, the emergence of hazard-specific degree programs, and shifting university-based programs from those that are theoretical and discipline specific to interdisciplinary programs targeted at solving social problems. These changes are all indicative of a research future that is more likely to provide the interdisciplinary work needed to support sustainable hazards mitigation.

Sustainable hazards mitigation is an alternative approach to dealing with hazards and disasters. Domestic model communities, international research comparisons, and alternative approaches to sharing knowledge with other nations all should hold prominent places in America's future.

CHAPTER NINE

Getting from Here to There

NATURAL HAZARDS MITIGATION will not be successful at reducing losses and disruption in the long term until it is integrated into the considerations of the daily activities of everyone who has an influence on future losses. This, in turn, will not be possible until hazards mitigation is housed within a redesigned national culture that favors sustainable development and people are reorganized to support that cultural shift. This book's proposal is really a double challenge: to promote greater acceptance of sustainable development in the United States and throughout the globe and to commit to making hazard mitigation and disaster management consistent with it.

This concluding chapter recommends ways to implement that shift, recommends actions to reduce disaster losses consistent with that shift, and points to the sustainable hazards mitigation research that is needed to acquire the knowledge to fuel general as well as specific progress. (Research recommendations in traditional hazards areas can be found in Appendix A.) Some of the recommendations seem like old news, since they echo the words of those who came before. But they are reiterated here because they still have not been met. Other recommendations will seem redundant because the nation has already begun moving down some of these paths.

ONE VIEW OF SUSTAINABLE HAZARDS MITIGATION IN ACTION

To bring together the disparate elements of sustainable hazards mitigation discussed in earlier chapters, this section presents an idealized view of how such mitigation might function. It should be emphasized that this is only one idea and is meant merely to illustrate how the components could fit together.

Striving for Local Consensus and a Common Agenda

All stakeholders in a community need to be brought to the point of taking responsibility for recognizing their locale's environmental resources and the environmental hazards to which it is prone. Stakeholders should use a consensus-building approach to determine community goals for the principles of sustainability: quality of life, disaster resiliency, economic vitality, environmental quality, and inter- and intragenerational equity.[1]

Consensus on the components of sustainability could be built around answers to questions such as: (1) What kind of lifestyle do people want and need? (2) In what manner should people live now so that future generations are not penalized? (3) How many future generations should be taken into account? (4) What is essential for a high quality of life? (5) At what point should growth be limited? (6) What risks are people willing to take in their interactions with the environment? (7) Under what circumstances should prevention be chosen versus permissible hazard? (8) What resources should be sustained and for whom? (9) Who or what will monitor and manage continued sustainability and in whose interest?

A second major step would be a local determination of disaster resiliency—the amount and kind of future losses that the stakeholders decide is locally tolerable (i.e., that which could be borne solely by those who experience them). Each locality would have a different vision of its indigenous tolerance for future disaster losses in terms of, for example, lives

[1]A stakeholder is any person or group that can affect or be affected by one or more of the issues of sustainability, and could differ from issue to issue.

lost, injuries sustained, people rendered homeless, interrupted critical facilities and lifelines, transportation system damage, and dollars lost to damage in other sectors of the constructed environment and economic systems.

Through this process a locality comes to realize that it controls the character of its future disasters. This would empower the local citizenry for hazards mitigation in two ways. First, it would install the unavoidable point of view that decisions made today determine what losses are experienced in the future. Second, it would help localities consider disaster losses and mitigation in the long-enough term to include future disasters that can go unnoticed when shorter time horizons are operative. It should be noted that localities will have or get substantial control over their futures but that there will be many externalities they will have to recognize as well. It is important to note that full consensus may never be reached and may not even be desirable. What is important is that the process be engaged in, for the information it generates and distributes, for the sense of community it can foster, for the ideas that grow out of it, and for the sense of ownership it creates.

Under the consensus-building approach, each stakeholder's policies, programs, work plans, and subsequent actions will be consistent with a shared common vision for the future. It will result in a holistic approach because hazards mitigation will be considered in the context of and integrated into the other worthwhile community goals that have historically served to constrain local hazards mitigation actions.

Building Networks

A way of organizing is needed to both forge the local consensus and nurture it through constant change and the challenges of implementation. The organizational mode must be flexible so that participating actors can be shifted to accommodate changes in knowledge and in the natural and human systems that determine hazard.

One workable structure may be networks of community stakeholders who would engage in collaborative problem solving—sustainable hazards mitigation networks. These sorts of flexible and collaborative intergovernmental forms have been used in other environmental management issues. Under this suggested structure, federal and state agencies would provide the cooperative leadership and support needed to bring these local networks into existence and to ensure they are successful. The areas covered by the networks would not necessarily correspond to an existing political entity but would follow natural boundaries that pertain to the hazard or resource at issue: watershed, earthquake zone, and so forth.

Qualities of Networks

Sustainable hazards mitigation networks should possess the qualities needed to integrate member goals and approaches to hazard mitigation in a holistic way. First, they would have a flat hierarchy to foster a sense of teamwork. Second, membership would be open to all, to enable the involvement of all areas of knowledge, different types of expertise, and all points of view. Third, the partners in the networks would use a consensus approach to determine solutions to problems, since "majority rule" can result in a disaffected minority. Fourth, networks would be driven by a common vision and set of values. Fifth, partners in the network would emphasize ongoing learning and continuous decisionmaking. Sixth, networks would evaluate their performance continuously and change accordingly. Seventh, network partners would refrain from exploiting their expertise to gain political leverage.

Network Membership

Sustainable hazards mitigation networks would be made up of the wide range of actors needed to address hazards at the local level in comprehensive and integrated ways. They would also include those people needed to merge hazards mitigation with the other goals of sustainability—environmental quality, economic vitality, inter- and intragenerational equity, and quality of life. Networks should be erected from the bottom up, that is, beginning at the local level where decisions about the future are actually made.

The formation of sustainable hazards mitigation networks might well begin with the state hazards mitigation officers, who, if they have not already done so, will form working relationships with natural resource managers, planners, elements of the private sector, nongovernmental organizations, expert advisers, state and federal science and policy agencies, and representatives of private citizens. Federal agencies could foster and support the process; provide technical assistance and access to experts and knowledge; and provide oversight for integrating local actions into state, national, and global activities.

Paying for Networks

Sustainable hazards mitigation networks would need to be externally funded so that participation in a network did not require a financial contribution from the individual partners—but they would be economical. Network partners could share limited resources, innovations, and expertise. Since there would be fewer networks than there are stakeholder groups today, national-level breakthroughs in knowledge and technological innovations could be disseminated more effectively and economically than at present.

A Nonlinear Approach

Networks should be equipped with whatever they need to redesign themselves and change their activities quickly in response to changing problems. First, there would be ground rules to enable the independent actors in a network to take integrated steps; form a system of interdependent activities; and coordinate their ideas, approaches, policies, strategies, and operations. Every stakeholder in the network would be considered vital to achieving the vision. Second, partners would appreciate and have access to the emerging trends and issues at higher levels—state, national, and global—that affect and are affected by their actions. Third, networks should have support in terms of resources, political access, legal authority, research findings, information sharing, and technical assistance. Last, networks would be empowered to innovate, for example, by sharing the beliefs that sustainable mitigation is in their hands, that it can only be carried forward through local innovation, and that state and federal leaders trust them to determine their own future.

Tools to Support Network Decisionmaking

It is difficult to anticipate everything that sustainable hazards mitigation networks might need to make informed, holistic, and state-of-the-art decisions, but at a minimum they would need access to local risk assessments, computer-generated decision aids, and holistic government policies.

Local Risk Assessments

Risk assessment combines information on a physical hazard (hazard assessment) with information on likely losses (vulnerability assessment) to estimate future disasters. Integrated multihazard community-scale risk assessment maps are not available for the majority of communities in the United States, yet these would be essential for every geographical area under the decisionmaking authority of each sustainable hazards mitigation network. These integrated assessments should do more than estimate what is in place today; they should also estimate future changes in risk from hazards and in environmental quality, based on current estimates of regional and global processes that could alter them in the future, such as climate change or changes in the size, composition, and distribution of population. Such risk assessments must be updated periodically to account for physical and societal changes.

Deterministic and probabilistic maps of potential disaster agents can be constructed at national, regional, urban, and site-specific scales for various scenarios and exposure times, but most such existing maps are for the national or state levels. Although progress has been made toward assessing the vulnerability of buildings, lifelines, and natural resources to various kinds of natural hazards, vulnerability assessment lags hazards assessment in the United States. Some data do exist about the potential for structural impacts and health impacts for various classes of buildings and infrastructure.

Computer-Generated Aids

Computer-generated models can estimate the likelihood of various loss levels in different geographic regions. This type of knowledge about future hazards would be needed by sustainable hazards mitigation networks to support their decisionmaking. The models in use today are sophisticated and include data on hundreds of variables. Historical data are collected on each of the fundamental components of a particular hazard; the data are analyzed to develop descriptions of the hazard's behavior; and the descriptions are programmed into the model as a basis for simulation. Other geophysical information is also included, such as detailed data on elevation, terrain conditions, and soil types. Models also require specific information on structures at risk, property values, construction types, and other engineering and architectural features. Ideally, future models would estimate with reasonable accuracy what today's models offer but would also project many other variables. The models would enable network decisionmakers to "see" the community-of-the-future consequences of every decision they make today.

Holistic Government Policies

Most of the hazards-related state and federal laws and policies in place today are an amalgamation of separate well-intentioned ideas that sprouted at different points in history. They are largely unrelated, each focusing on narrow program objectives and measurable individual goals, and most are based on the perspective that a narrow slice of mitigation is good without considering how separate actions may interact to affect actual future hazards. A holistic view of hazards and risk is missing, yet local networks cannot be expected to function effectively without better guidance and a more consistent framework.

Measuring Local Progress

Progress toward sustainable hazards mitigation should be measured. Gauging progress would be problematic even if sound indicators were in hand because baseline measurements of hazard and vulnerability do not exist for most communities. One strategy for overcoming this would be for each sustainable mitigation network's vision of its local tolerance for future disaster losses to be expressed in terms of lives lost, injuries sustained, people left homeless, environmental damage, interrupted critical facilities and lifelines, transportation system damage, damage in other sectors of the constructed environment and economic systems, and so forth. Future disasters would be modeled to determine anticipated losses in the specified categories. The gap between the two estimated sets of losses—tolerable indigenous losses versus likely future losses—would constitute the "mitigation gap" that local networks would work to close.

This method would also be used to simultaneously estimate progress toward the locally desired levels of environmental quality, inter- and intragenerational equity, quality of life, and achieving a sustainable local economy. The effectiveness of local networks would be measured multivariately in terms of progress toward these five objectives. Progress toward achieving sustainability could not be compared across communities in absolute terms because progress would be relative to a local setting's hazard and its own tolerance for future losses. However, it could probably be aggregated in a rough way for regional, state, and national levels.

Making Broader Progress

This idealized scheme could be implemented gradually. Initially, local networks could augment the existing decisionmaking framework. Then, gradually, reprogramming could take place as consensus dictates, experience grows, and needed tools become more refined and more widely available. It should be noted that achieving individual sustainable communities across the nation is still a rather narrow goal. A series of sustainable communities may not add up to a sustainable nation or a sustainable globe. There are still many factors to be addressed, such as economic externalities. Larger units of complexity must also eventually be addressed. But working from the bottom up arguably is the best place to start.

SOME POLICY AND RESEARCH STEPS THAT NEED TO BE TAKEN

The shift to a sustainable approach to hazards mitigation will require some extraordinary actions, and some research is needed to inform and evaluate those actions. Many of the initial steps have already been taken, and there are many entities already at work in the nation to expand thinking about the sorts of issues that will eventually contribute to sustainable mitigation. The hazards community needs to take some steps on its own, but it also needs to find and join forces with initiatives that are already under way.

A change in thinking is needed—from the short to the long term, from within singular disciplines and solutions to across hazards and disciplines. Existing constraints in current forms of organization need to be overcome. Moreover, sustainable hazards mitigation requires that local communities forge consensus and a common agenda for their future across diverse stakeholders; building and empowering networks of local actors; a nonlinear approach that can deliver to local decisionmakers new and diverse information as it is learned, when it is needed, and in ways that can be useful; the development of tools to support local decisionmaking, such as comprehensive risk assessments and computer-based decision support systems; holistic government policies that do not work at cross-purposes; and an open and effective approach to measuring progress. Some of the most significant steps that need to be taken are described below and research is needed to support moving forward with those actions and then evaluating their outcomes and effectiveness.

Consolidating Existing Knowledge and Putting It into Practice

This recommendation involves spreading the word about sustainable mitigation, making further improvements in disseminating what is known about hazards to the appropriate players, and improving existing practices by being sure that what is known actually gets implemented. Crucial steps for doing this are described below.

Sharing Knowledge About Hazards and Sustainability

Although popular understanding of immediate risk has increased enormously since the nation's first assessment of natural hazards (White and Haas, 1975), understanding of long-term risk is more limited. First and foremost is the need for information about risk, not just about hazards. State and local policymakers and citizens need to understand, for example, not only that a particular area is subject to seismic or flood hazards but also that, with existing or proposed use of the area, certain levels of loss can be expected and those losses will have certain undesirable consequences—particularly if existing federal subsidies are no longer available. This information is not now available and is not likely to become available without investment in research to develop and disseminate risk analysis tools to local governments.

Conduct National Forums on Sustainability

To shift the nation toward sustainability, leadership is needed from federal agencies and interagency subgroupings charged with developing hazards policy and/or conducting hazards research. Agencies like the Federal Emergency Management Agency (FEMA), the U.S. Geological Survey (USGS), the National Oceanic and Atmospheric Administration (NOAA), the U.S. Environmental Protection Agency, the U.S. Forest Service, and the Subcommittees on Natural Disaster Reduction and Risk Analysis of the Committee on the Environment in the Office of Science and Technology Policy all have agency-specific vision statements that foretell sustainable hazards mitigation. These agencies should generate forums that bring hazards experts together to discuss sustainable hazards mitigation and generate alignment and partnerships for bringing it into existence.

A variety of interdisciplinary and across-hazards forums already in place (workshops, conferences, meetings) could be used for this purpose—for example, the annual Natural Hazards Workshop sponsored by the Natural Hazards Research and Applications Information Center of the University of Colorado, the annual meeting of the International Association of Emergency Managers (formerly the National Coordinating Council on Emergency Management), and many others. Consequently, the national dialogue could begin at little or no additional cost.

Establishing a Loss Data Archive

The nation must devote resources to collect, analyze, and store hazard and loss data gathered on actual disasters. Data on the type of loss, location, specific cause, and actual dollar cost need to be compiled in a uniform fashion for across-hazards comparisons. Spatially related indicators derived from geographic information systems or digital map systems using satellite imaging could help collect the needed data. The archive should be accessible to the public. This is fundamental to the development of more sophisticated computer models capable of "predicting" future local vulnerability to disasters under a variety of conditions and projecting other aspects of sustainability, such as environmental quality, economic vitality, and social equity.

Developing Support Tools for Locals

The kind of risk and vulnerability knowledge needed to foster holistic local decisionmaking is not yet fully in hand and what is available is not sufficiently integrated, but the direction is clear, and several immediate steps will make progress possible.

Integrated multi-hazard, consistently scaled (preferably 1 inch = 2,000 feet or better) risk assessment and vulnerability maps are not available for the majority of communities in the United States, yet these would be essential for sustainable hazards mitigation networks. Moreover, reasonably accurate local risk assessments for natural hazards would have to be combined with similar assessments of environmental quality and all other mappable aspects of sustainability. They would be most useful if they were to also estimate changes in risk from natural hazards and environmental quality based on current estimates of regional and global processes that could alter them in the future, such as climate change or changes in the size, composition, and distribution of populations. Interdisciplinary research to produce the information needed for local risk and vulnerability assessment should be given high priority.

Computer-generated models would be essential to estimate the likelihood of various loss levels at different locales. The models would need to include data on hundreds of variables, such as historical data on each hazard, elevation, terrain conditions, soil types, structures at risk, property values, construction types, and other engineering and architectural features. Future models must go beyond today's capabilities to project alternative levels of vulnerability based on future population growth and other factors; losses in future disasters based on alternative mitigation decisions made today, such as different land-use and building code decisions; and impacts on and changes in other aspects of sustainability, such as environmental quality, economic vitality, and social equity. Research to advance the development of computer decisionmaking aids like these is needed.

Providing for Holistic Education and Training

A variety of initiatives have been proposed or begun to expand education for hazards mitigation and emergency preparedness, and these should be continued. Two more initiatives are needed to support sustainable hazards mitigation. First, the holistic perspectives needed for sustainable hazards mitigation cannot be gained through education in just any one of the specialized fields available today. Members of the next generation must be educated across the disciplines that have knowledge applicable to sustainable hazards mitigation. Thus, interdisciplinary degree programs are needed. Second, although the number of hazards researchers has increased substantially over the past 20 years, their absolute number remains small. This group should collaborate to expand cooperative research; enhance the sophistication and breadth of the research produced; encourage standardization of instrumentation, data processing and documentation, and analytical approaches; and increase interactions among researchers.

Hazard and disaster researchers need to be provided with mechanisms, incentives, and reasons for transcending institutional boundaries and their own self-interests. The overhead charges associated with extramural funding when multiple institutions or investigators are involved need to be reduced. It should be possible to designate coinvestigators rather than a sole "principal." Procedures could be established to encourage graduate students to be hired as research assistants outside their home department or institution. It should be possible to make at least part of the grant funds for graduate students in the hazards field "portable." There are already models for crossing these barriers—in undergraduate exchanges among U.S. universities, in the long-standing junior-year abroad programs, and in the Senior Service colleges run by the military.

Third, training materials and courses should be improved for professionals in land-use planning, civil engineering and public works, building design and code enforcement, public finance, and emergency services. Such professionals should also be encouraged to develop their own information about hazards. The federal government has taken important steps in working with a number of relevant groups to expand knowledge (e.g., wind and seismic engineering, city and regional planning) that should be continued and expanded. Finally, knowledge about retrofitting existing buildings to reduce their vulnerability to various hazards must be diffused among the building trades and do-it-yourself enthusiasts.

Setting Up a Social Science Data Archive

A central repository for social science research data on hazards and disasters should be established. Researchers should be asked to file copies of data collection instruments, specifications of how and why the data were collected, and copies of proposals with the original data so that they are permanently available to other researchers. Such an archive could be expanded over time to include all related datasets and could even be a resource for practitioners and policymakers. It also might be used to encourage governments to standardize and centralize their data collections. The information in the repository should be maintained in a digitized format.

Establishing Holistic Government Policies and Frameworks

A holistic view of hazards and risk is missing from most of the hazards-related state and federal laws and policies in place today. Nor are the existing programs coordinated well enough to enable a holistic approach even when one is desired. A conference or series of conferences should be conducted at which federal, state, county, and city officials would reexamine the statutory and regulatory foundations of mitigation and preparedness for every hazard. There would be two goals: (1) to integrate and render consistent all policies and programs related to hazards and sustainability across all levels of government in the United States and (2) to build the principles of sustainable mitigation into such laws and programs. As one example, governments must do a better job of coordinating and integrating hazards policies with those for economic, social, and environmental goals. Furthermore, building the principles of sustainable mitigation into existing legislation should be a conscious aspect of reauthorization of ongoing programs.

Most government efforts to cope with hazards today are fragmented horizontally at each level of government, vertically between levels of government, and across different types of hazards. This dispersal makes it extremely difficult for local governments to deal with hazards in a coherent way. An additional consequence is that each division fails to recognize the integrated character of problems—especially environmental ones. Sustainable hazards mitigation will require more interaction across government agency boundaries and levels and across contemporary distinctions between academic disciplines, government, the private sector, and the public. A new approach must be forged that will take a holistic perspective of the complex problems surrounding hazards and the risks they engender, enabling people not only to design better solutions but also to view the impacts of interactions between mitigation activities themselves. Different authorities from different entities must be reassembled in new ways. People will have to begin working together in ever-changing ways that contradict current organizational structures.

Moreover, a number of possible measures of holistic government policies should be considered by a range of reasonable groups to bring a divergence of perspectives and viewpoints forward. Illustrative approaches include a joint congressional committee hearing; a report by the Congressional Research Service; a report by the White House Council on Sustainable Development; a national conference by the American Planning Association reviewing experiences in sample communities and areas; and a joint meeting of concerned federal, state, and national research organizations.

Fostering the Shift to Local and Regional Responsibility and Capability

Lack of state and local government commitment and capacity has been the Achilles heel of many previous efforts to deal with natural hazards. A central role of the federal government must be to foster collaborative state and local planning for sustainable mitigation. The new process would encourage all levels of government to define for themselves the meaning of sustainability and ways in which it can be accomplished. The federal government would lead by example and provide technical and financial assistance.

As this volume has demonstrated, the United States is rich with knowledge, expertise, information transfer organizations, government agencies with hazards-directed missions, research centers, professional associations, advocacy groups, and hazards-related businesses. These actors need to be integrated across hazards and specialties as well as aligned with groups concerned with related issues, such as natural resources, environmental quality, economic development, and quality of life. That knowledge and those people need to be directly accessible to localities. Three avenues are suggested for getting started on this: creating local sustainable hazards mitigation networks, setting up prototype regional projects to provide knowledge and expertise to those networks, and establishing model communities to illustrate sustainability. Research and policy analyses are needed to determine reasonable ways to shift to local and regional responsibility.

Developing Prototype Local Networks

Local stakeholders should be organized into flexible sustainable hazards mitigation networks that can adjust to rapid changes in knowledge and in the natural, human, and constructed systems that determine risk and vulnerability. This would show how the sustainable hazards mitigation networks can work and would establish their viability and credibility. Agencies such as FEMA, NOAA, the USGS, and state offices with emergency, geological, and climatological functions could provide cooperative leadership and support to develop these networks (both directly and through the regional outreach projects described below). These networks would use a consensus approach to develop a common vision and set of goals. They would approach hazard mitigation in a holistic way. Prototype local networks should be researched and evaluated to determine what works and what does not.

Developing Prototype Regional Projects

FEMA, the USGS, NOAA, and other federal agencies with outreach missions could create and support a series of prototype regional outreach projects. These projects should be independent of established government agencies and staffed with people trained in the physical and social sciences and engineering. The projects should focus on all of the hazards issues in their geographical areas, have access to the nation's experts, help establish local sustainable hazards mitigation networks and provide technical assistance to them, disseminate knowledge in useable forms to network partners, develop local interest in sustainable hazards mitigation, and produce model products that could be transferred to networks in other areas. These regional projects should be researched and evaluated to determine how to enhance their future effectiveness.

Approaches like this have yielded some of the most successful accomplishments in hazards mitigation since the nation's first assessment. Two examples are the Southern California Earthquake Preparedness Project and the Bay Area Regional Earthquake Preparedness Project, which were established by FEMA and the state of California in the 1980s to provide technical information and assistance to local governments and organizations. They brought together diverse experts, agencies, organizations, and experience to cooperate with other public- and private-sector parties in working on earthquake hazard mitigation and preparedness. Both projects were enormously successful.

Regional land-use planning has been advocated in the past for management of ecosystems, earthquake hazards, and flood hazards. Properly designed regions with jurisdictional boundaries that are congruent to natural boundaries provide a basis for managing natural resources and environmental conditions at an appropriate scale. A key impediment to the formation of regional tiers of government is that they pose potential threats to other levels of government and their constituencies. Furthermore, establishing meaningful "boundaries" for a multihazards approach will be a challenge.

The major successes in initiating regional planning and management of land use have involved creation by federal or state legislatures of new special-purpose organizations with the authority to implement their plans and policies. Regional partnerships between levels of government with less intervention-oriented objectives also have been successful. Prominent examples of regional planning and regulatory entities include the Adirondack Park Agency, the New Jersey Pinelands Commission, the Tennessee Valley Authority, Florida's water management districts, the Delaware River Basin Commission, and flood control agencies such as Denver's Urban Drainage and Flood Control District.

Developing Model Communities

Examples of how a community might look and function if it redesigned itself to become disaster resilient would help further a shift in public thinking about hazards. The Institute for Business and Home Safety's Showcase Communities Project and FEMA's Project Impact are two among other nationwide initiatives to help communities work toward disaster resiliency and sustainability. These projects feature public-private partnerships geared toward specific aspects of local disaster resilience. More of these efforts should be undertaken, and involved communities should be extensively researched to determine the factors that influenced their successes and failures from social scientific, natural scientific, and engineering viewpoints.

Making Changes in Public Policies

For purposes of hazards mitigation, a sensible approach for the federal government would feature cooperation and collaboration with state and local governments while also being consistent with ending subsidies; enhancing information; fostering consensus and commitment; building state, local, and private capacity; and acting in ways consistent with state and local plans and policies. Policies should enhance local government interest and ability to work toward policy goals. Federal programs may prescribe planning or procedures but not the specific means for achieving desired outcomes. Financial and technical assistance could both enhance local commitment and increase its capacity.

Limiting the Subsidization of Risk

As long as risk is subsidized (or people believe it is), it will be difficult to convince the public of the need to take responsibility for decisions that put them and their property at risk. The various forms of federal subsidies and disaster relief should be examined (e.g., shoreline protection, flood control, tax write-offs). Detailed policy analyses of alterative ways to reduce or eliminate subsidization of risk are needed before action is taken; then interdisciplinary research into the actual impacts of any policies adopted should be conducted.

Most, if not all, of the costs of hazard control structures should be covered by direct state, local, and private beneficiaries. Federal appropriations to the agencies providing these services could be phased out, and they could be required to operate on a full cost recovery basis. States and localities could pass costs back to the owners and occupants of hazardous areas through mechanisms such as benefit assessment districts and impact fees and taxes. Most of these changes could be made with minimal additional research or adverse effect, although research would be needed to develop cost recovery mechanisms and estimate the likely economic consequences for people and property currently at risk.

Making Better Use of Incentives

Virtually all federal aid for infrastructure and economic development should be dependent on community participation in land use and other planning since natural hazards, if ignored, have the potential to ruin such federal investments. Minimum loss standards could be set, below which communities and states could not fall without losing federal aid.

A federal hazardous areas management (or similar) act would stimulate planning and management to accomplish a balanced mix of economic development and environmental goals. Such an act would not impose national standards on state and local governments but would require them to initiate planning and management processes so that exposure to all natural hazards is considered in public and private development and redevelopment. Any future legislation should include a mechanism to assure consistency both horizontally (between states that share a hazard or resource, for example) and vertically (among the federal, state, and local levels).

FEMA could require local preparation of floodplain management plans, as already authorized by Congress, as a condition for continued participation in the National Flood Insurance Program (NFIP). Such plans already are credited under the NFIP's Community Rating System, which rewards communities that are going beyond the minimum federal standards. The federal government could require areawide hazard adjustment plans as a condition of federal assistance with such measures as flood control structures and beach nourishment. This approach should be researched and evaluated over time to determine its impact and effectiveness. A requirement for periodic interdisciplinary evaluations should be written into the initial legislation.

Bringing in the Private Sector and Nonprofit Groups

Much has been done in the past decade, especially very recently, to integrate public- and private-sector activities related to hazards mitigation. But more needs to be done not only to continue building public-private partnerships for specific activities but also to allow ongoing programs to evolve toward better use of both sectors. There should be wider participation by the public, government, and the construction and insurance industries in the continuous process of improving building codes. Complete and candid clarification of cost versus safety and other criteria would benefit everyone involved and promote better understanding and enforcement of the codes.

There is significant potential for linking insurance to the adoption of mitigation measures that reduce damage to existing buildings and the amount of new development in hazardous areas. The industry is beginning to take steps in this direction. The Insurance Services Office and the Institute for Business and Home Safety have begun to establish a building code effectiveness grading schedule to rate code enforcement at the local level. Insurance availability and rates would be partially determined by a community's grade. Florida has already enacted legislation that allows insurance companies to include this in their rate factors.

Minimum levels of damage resistance should be incorporated into building codes and verified by inspections. Structures that meet the requirements could be given a "seal of approval" that could be recognized and relied on by insurers, regulators, financial institutions, and property owners. Insurers, for example, could use the inspection data in deciding whether to provide insurance and what rates to charge. Such a program would likely lead to well-enforced building codes. It also would provide incentives for mitigation, produce more equitable insurance premiums, and give regulators a better measure of the risk of catastrophe faced by insurers doing business in their states. The impacts of actions such as these should be investigated to determine their effectiveness and their contributions to sustainable hazards mitigation.

Measuring Progress of Past and Future Efforts

It is not enough to set a direction or establish a program and then proceed more or less on faith. A better job must be done now and in the future of setting interim goals, determining how progress toward them is to be measured, and keeping accurate data that contribute toward that understanding. Part of this effort must be integrated with a national database, as discussed above. Part of it must involve a commitment to incorporate continuous evaluation into all future mitigation efforts.

Evaluation of Programs Already in Place

It has been 30 years since the NFIP began its combined program of incentives, land-use and building standards, insurance, and federal-state-local and private/public-sector coordination. It is the nation's broadest and arguably best-designed mitigation program, and many of its components parallel the sort of integrated approach advocated by practitioners and researchers today. Yet no thorough evaluation of its effectiveness has ever been conducted. It would be a big project that may require millions of dollars if it is to be done accurately, but it is absolutely vital if Congress is to make any informed decisions about insurance for other hazards. And the information yielded by such an evaluation would surely inform other aspects of long-term mitigation and sustainability. The NFIP is only one example of programs in place that are proceeding without anyone having a clear idea of what their real costs and benefits are. Hard questions should be asked and then investigated regarding all of the nation's mitigation schemes so that lessons can be learned, mistakes avoided, and money saved.

A National Repository of Costs and Losses

No one really knows what the nation and its citizens lose from natural and technological hazards and disasters each year. A comprehensive database that contains information about current levels of vulnerability to natural hazards on national and local scales, compilations of past losses (see discussion above) and costs (both direct and indirect), and catalogs of preevent mitigation activities and expenditures is needed. The data should be standardized and collected systematically. Furthermore, there must be public access to the data so that everyone has the opportunity to make better informed decisions.

Future Baselines for Sustainability

One strategy for measuring the progress of sustainable mitigation would be for each mitigation network's vision of its local tolerance for future disaster losses to be expressed in terms of lives lost, injuries, environmental damage, interrupted critical facilities and lifelines, transportation system damage, and the like. Computer models of future disasters could be run to determine anticipated losses in the specified categories.

The gap between the two estimated sets of losses—tolerable indigenous losses versus likely future losses—would constitute the "mitigation gap" that local networks would work to close. This method would also be used to simultaneously estimate progress toward the locally desired levels of environmental quality, inter- and intragenerational equity, quality of life, and achieving a sustainable local economy. The effectiveness of local networks would be measured multivariately in terms of progress toward these five objectives.

Transferring Knowledge Internationally

The United States must share knowledge and technology for sustainable hazards mitigation with other nations and be willing to learn from others as well. Disaster and development experts should work together to address the root causes of disaster vulnerability, be they overgrazing, deforestation, unplanned development, or poverty. Disaster aid should not come in the form of increased debt. Finally, there needs to be a general move toward integration of work being done. Disaster reduction should be a part of everyday development processes.

The West should be learning from countries that have managed to improve quality of life without significantly increasing their consumption of natural resources. Local traditions and knowledge of past disasters should be seen as useful input into the planning or relief process. At the same time, other countries in the West have had experiences that would benefit the United States. For example, New Zealand has received worldwide attention for its reform of resource and environmental management and its vision of integrated "effects-based" environmental management. The intergovernmental regime in New South Wales provides a noteworthy example of a shift from a heavy hand to a more flexible approach to help local governments cope with flood hazards.

U.S. agencies have historically been constrained to concentrate on hazards in this country, although they have participated in international collaborative projects and engaged in scientific communication. New technologies for information dissemination and participation in the International Decade for Natural Disaster Reduction should make it easier for the United States to more fully engage in knowledge transfer throughout the world.

Determining the Hazardousness of the Nation

The hazardousness of the United States is changing, but it is not really known what disaster risks lie ahead. There is inadequate information about the baselines of, and changes in, the environmental, sociodemographic, and constructed systems of the nation that are shaping future losses.

A national assessment of risks from all hazards is needed. This research effort should blend a comprehensive treatment of the interaction of and changes in the environmental systems that produce extreme events, the people and communities that experience those events, and the constructed environment that is affected. The assessment should support and complement the local risk and vulnerability assessments called for previously in support of local sustainable hazards mitigation networks. It should cover the largest relevant global physical processes and changes down through their hazards and implications for the people, resources, buildings, structures, and decisionmaking in local communities. The assessment should not stop at collecting the more obvious data but should also include the way in which disaster vulnerability is distributed; how various forms of diversity (ethnicity, gender, socioeconomic status, physical challenges) affect vulnerability; public and private mechanisms for spreading these risks; and attention to the special situations of rural areas.

A FINAL WORD

This book is about new possibilities in the ways in which the United States thinks about and approaches hazards, disasters, and the natural environment. The nation needs to acknowledge that it will never be totally safe from disasters; it needs to eliminate policies, actions, and rhetoric that lure citizens into a false sense of security; and it should not view disasters as problems that can be solved in isolation but rather as symptoms of broader and in many ways unaddressed problems. Members of the hazards community—researchers, policymakers, and users alike—need to acknowledge that the actions they take today are two sided: today's actions to deal with vulnerability in the short term may unintentionally redistribute and postpone losses into an ever more catastrophic future.

Capitalizing on the knowledge that has been accumulated and progress made to date and then moving toward a vision of sustainability will require a cooperative approach that integrates expertise across professions. Ideas, criticisms, and different points of view will combine to change direction toward a course that will create a more sensible future. In order for the concept to become part of daily life, shifts will be needed in the values, practices, and approaches that guide interactions between people and the environment. A broader view of the causes of natural disaster losses is needed. Long-term perspectives must be added to short-term interests. At the very least, a nationwide conversation should begin to consider actions that will link hazards mitigation and disaster response to the broader goals of sustainability. What is important is that people unite in a common commitment to sustainable mitigation and begin to adopt attitudes and practices that will make it so.

References

Ahrens, D. 1993. *Essentials of Meteorology*. St. Paul, Minn.: West Publishing Co.

Anderson, L. R., J. R. Keaton, T. Saarinen, and W. G. Wells II. 1984. *The Utah Landslides, Debris Flows, and Floods of May-June 1983*. Washington, D.C.: National Academy Press.

Anderson, M., and P. Woodrow. 1989. *Rising from the Ashes: Development Strategies in Times of Disaster*. Boulder, Colo.: Westview Press.

Andrews, P. M. 1988. The National Fire Danger Rating System as an Indicator of Fire Business. Pp. 49-56 in *Proceedings of the Ninth Conference on Fire and Forest Meteorology*, San Diego, Calif. Boston: American Meteorological Society.

Atomic Safety and Licensing Board. 1986. In the matter of Carolina Power and Light Company, NRC Docket No. 50-400-0L. 23 NRC 294.

Baird, A., P. O'Keefe, K. Westgate, and B. Wisner. 1975. Towards an Explanation of Disaster Proneness. Disaster Research Unit Occasional Paper No. 10. Bradford, England: University of Bradford.

Bakun W. H., and A. G. Lindh. 1985. The Parkfield California, Earthquake Prediction Experiment. *Science* 229:619-624.

Beatley, T., D. J. Brower, and A. K. Schwab. 1994. *An Introduction to Coastal Zone Management*. Washington, D.C.: Island Press.

Berke, P. 1995. Natural Hazard Reduction and Sustainable Development: A Global Assessment. *Journal of Planning Literature* 9(4):370-382.

Berke, P. R., J. Kartez and D. Wenger. 1993. Recovery After Disaster: Achieving Sustainable Development, Mitigation and Equity. *Disasters* 17(2):93-109.

Blaikie, P., T. Cannon, I. Davis, and B. Wisner. 1994. *At Risk: Natural Hazards, People's Vulnerability, and Disasters*. New York: Routledge.

Bolt, B. 1993. *Earthquakes*. New York: W. H. Freeman.

Burby, R. J., and S. P. French with B. A. Cigler, E. J. Kaiser, D. H. Moreau, and B. Stiftel. 1985. *Flood Plain Land Use Management: A National Assessment*. Boulder, Colo.: Westview Press.

Burby, R. J., and P. J. May, 1997. *Making Governments Plan: State Experiments in Managing Land Use*. Baltimore: Johns Hopkins University Press.

Burby, R. J., B. A. Cigler, S. P. French, E. J. Kaiser, J. Kartez, D. Roenigk, D. Weist, and D. Whittington. 1991. *Sharing Environmental Risks, How to Control Governments' Losses in Natural Disasters*. Boulder, Colo.: Westview Press.

Burton, I., R. W. Kates, and G. F. White. 1978. *The Environment as Hazard*. New York: Oxford University Press.

Burton, I., R. W. Kates, and G. F. White. 1993. *The Environment as Hazard, Second Edition*. New York: Oxford University Press.

California Seismic Safety Commission. 1994. *Research and Implementation Plan for Earthquake Risk Reduction in California 1995 to 2000*. Sacramento, Calif.: California Seismic Safety Commission.

Carr, C. 1982. The Political Economy of Hunger and Desertification in Pastoral Ethiopia and the Challenge of Future Development. Paper presented at the annual meeting of the Association of American Geographers, San Antonio, April 25-28.

Central U.S. Earthquake Consortium. 1997. *Disaster Resistant Communities*. Memphis, Tenn.: CUSEC.

Christopherson, R. W. 1997. *Geosystems, An Introduction to Physical Geography*. Upper Saddle River, N.J.: Prentice-Hall.

Cochrane, H. C., M. J. Bowden, and R. W. Kates. 1974. *Social Science Perspectives on the Coming San Francisco Earthquake: Economic Impacts, Prediction, and Reconstruction*. Working Paper #25. Boulder: Institute of Behavioral Science, University of Colorado.

Cochrane, H. C., P. Laub, and J. Barth. 1992. Banking and Financial Markets. Chapter 4 in *Indirect Economic Consequences of a Catastrophic Earthquake*. Report prepared for the Federal Emergency Management Agency. Washington, D.C.: FEMA.

Committee on Science, Engineering, and Public Policy. 1992. *Policy Implications of Greenhouse Warming: Mitigation, Adaptation, and the Science Base*. Panel on Policy Implications of Greenhouse Warming, National Academy of Sciences, National Academy of Engineering, and Institute for Medicine. Washington, D.C.: National Academy Press.

Crosby, B. 1993. War in Our Forests. *Sunset* (July):63–69.

Cruden, D., and D. Varnes. 1996. Landslide Types and Processes. Chapter 3 in *Landslides: Investigation and Mitigation*. Special Report 247. Washington, D.C.: National Research Council, Transportation Research Board.

Cutter, S. L., and M. Ji. 1997. Trends in U.S. Hazardous Materials Transportation Spills. *Professional Geographer* 49(3): 318-331.

Dewey, J. 1929. The Quest for Certainty. In J. Boydston, ed. *The Later Works: 1925-1953*. Carbondale, Ill.: Southern Illinois University Press.

Dewey, J. 1938. *Logic, the Nature of Inquiry*. New York: Holt, Rinehart & Winston.

Drabek, T. E. 1986. *Human Systems Responses to Disaster*. New York: Springer-Verlag.

Drabek, T. E., and Gerard J. Hoetmer, eds. 1991. *Emergency Management: Principles and Practice for Local Government*. Washington, D.C.: International City Management Association.

Dynes, R. R., and T. E. Drabek. 1994. The Structure of Disaster Research: Its Policy and Disciplinary Implications. *International Journal of Mass Emergencies and Disasters* 12:5-23.

Eagleman, J. 1990. *Severe and Unusual Weather*. Lenexa, Kans: Trimedia Publishing Co.

Earthquake Research Institute. University of Tokyo, Japan. 1996. Information at ftp://ftp.eri.u-tokyo.ac.jp/‘

Ellsworth, W. L., and G. C. Beroza. 1995. Seismic Evidence for an Earthquake Nucleation Phase. *Science* 268:851.

Environmental Protection Agency. 1993. *A Guidebook to Comparing Risks and Setting Environmental Priorities*. EPA 230-B-93-003, Washington, D.C.: EPA.

Environmental Protection Agency. 1988. *Review of Emergency Systems: Final Report to Congress*. Washington, D.C.: EPA.

Federal Emergency Management Agency. 1991. Principal Threats Facing Communities and Local Emergency Management Coordinators. A Report to the U.S. Senate Committee on Appropriations. Washington, D.C.: FEMA.

Federal Emergency Management Agency. 1992. *Building Performance: Hurricane Andrew in Florida*. Washington, D.C.: U.S. Government Printing Office.

Federal Emergency Management Agency. 1995. *National Mitigation Strategy*. Washington, D.C.: U.S. Government Printing Office.

Fishbein, M., and M. Stasson. 1990. The Role of Desires, Self-Predictions, and Perceived Control in the Prediction of Training Session Attendance. *Journal of Applied Social Psychology* 20:173-198.

Friesma, H. P., J. Caporaso, G. Goldstein, R. Lineberry, and R. McCleary. 1979. *Aftermath: Communities After Natural Disasters*. Beverly Hills, Calif.: Sage.

Fritz, C. 1961. Disaster. Pp. 651-694 in *Contemporary Social Problems,* R.K. Merton and R.A. Nisbet, eds. New York: Harcourt Press.

Geis, D., and T. Kutzmark. 1995. Developing Sustainable Communities: The Future Is Now. *Public Management* 77(8):4-13.

Glickman, T. S., and D. Golding. 1992. Recent Trends in Major Natural Disasters and Industrial Accidents. *Resources for the Future Newsletter* 108:9-14.

Hays, W. W. 1981. *Facing Geologic and Hydrologic Hazards: Earth-Science Considerations*. Professional paper 1340-B. Reston, Va: U.S. Geological Survey.

Hays, W. W., and B. M. Rouhban. 1991. U.S. National Earthquake Program Technology Transfer. *Episodes* 14 (1):66-72.

Hewitt, K., 1976. Earthquake Hazards in the Mountains. *Natural History* (May): 30-37.

Hewitt, K., ed. 1983. *Interpretations of Calamity: From the Viewpoint of Human Ecology*. London: Allen and Unwin.

Hewitt, K. 1997. *Regions of Risk: A Geographical Introduction to Disasters*. Essex, U.K.: Longman, Ltd.

Hill, C. 1996. Mayday! *Weatherwise* 49(3):25–28.

Hillaker, H. J., Jr., and P. J. Waite. 1985. Crop-Hail Damage in the Midwest Corn Belt. *Journal of Climate and Applied Meteorology* 24:3–15.

Hird, J. A. 1994. *Superfund: The Political Economy of Environmental Risk*. Baltimore: Johns Hopkins University Press.

Hogarth, R. M., and H. Kunreuther. 1993. *Decisionmaking Under Ignorance: Arguing with Yourself*. Paper #93-10-04. Philadelphia: University of Pennsylvania, Risk Management and Decision Processes Center.

Horlick-Jones, T. 1995. Modern Disasters as Outrage and Betrayal. *International Journal of Mass Emergencies and Disasters* 13 (3):305-315.

Howe, C. W. and H. C. Cochrane. 1993. *Guidelines for the Uniform Definition, Identification, and Measurement of Economic Damages from Natural Hazard Events*. Special Publication #28. Boulder: Natural Hazards Information Center, University of Colorado.

Insurance Research Council. 1991. *Small Business Attitude Monitor*. 91 (July)

Insurance Research Council. 1996. *Public Attitude Monitor* 96 (January).

Insurance Research Council and Insurance Institute of Property Loss Reduction. 1995. *Coastal Exposure and Community Protection: Hurricane Andrew's Legacy*. Wheaton, Ill.: IRC.

Interagency Floodplain Management Review Committee. 1994. *Sharing the Challenge: Floodplain Management into the 21st Century*. Washington, D.C.: U.S. Government Printing Office.

Intergovernmental Panel on Climate Change. 1995. *Second Assessment Report*. Cambridge, Mass.: Cambridge University Press.

International Journal of Mass Emergencies and Disasters. 1995. Special Issue: What Is a Disaster? Six Views of the Problem. 13(3).

Kates, R. W. 1962. *Hazard and Choice Perception in Floodplain Management*. Research Paper No. 78. Chicago: University of Chicago Press.

Kreps, G. A. 1991. Organizing for Emergency Management. Pp. 30-54 in *Emergency Management: Principles and Practice for Local Government,* T. E. Drabek and G. J. Hoetmer, eds. Washington, D.C.: International City Management Association.

Kreps, G. A., and T. E. Drabek. 1996. Disasters Are Nonroutine Social Problems. *International Journal of Mass Emergencies and Disasters* 14:129-153.

Kunreuther, H., and P. Slovic, eds. 1996. Challenges in Risk Assessment and Risk Management. *Annals of the American Academy of Political and Social Science* 545:1-220.

Kunreuther, H., R. Ginsberg, L. Miller, P. Sagi, P. Slovic, B. Borkan, and N. Katz. 1978. *Disaster Insurance Protection: Public Policy Lessons*. New York: Wiley.

Lindell, M., and R. Perry. 1992. *Behavioral Foundations of Community Emergency Planning*. Washington, D.C.: Hemisphere.

MacIver, R. M. 1931. *Society: Its Structure and Changes*. New York: Long and Smith.

McAdie, C. J. 1996. Operational Trends in Hurricane Forecasting. Paper presented at the 18th Annual National Hurricane Conference, Orlando, Fla.

Marston, S. 1982. *A Political Economy Approach to Hazard: A Case Study of California Lenders and the Earthquake Threat*. Natural Hazards Research Working Paper No. 49. Boulder: University of Colorado, Institute of Behavioral Science.

Maunder, W. J. 1986. *The Human Impact of Climate Uncertainty: Weather Information, Economic Planning and Business Management*. London: Routledge Press.

May, P. J. 1985. *Recovery from Catastrophes: Federal Disaster Relief Policy and Politics*. Westport, Conn.: Greenwood Press.

May, P. J., R. J. Burby, J. Dixon, N. Ericksen, J. Handmer, S. Michaels, and D. I. Smith. 1996. *Environmental Management and Governance: Intergovernmental Approaches to Hazards and Sustainability*. London: Routledge Press.

May, P. J., and W. Williams. 1986. *Disaster Policy Implementation: Managing Programs Under Shared Governance*. New York: Plenum.

Medvedev, G. 1990. *The Truth About Chernobyl*. New York: Basic Books.

Miami Herald. 1992. Special Report: What Went Wrong. Dec. 20.

Micklin, M. 1973. *Population, Environment, and Social Organization*. Hinsdale, Ill.: Dryden Press.

Mileti, D. S., and C. Fitzpatrick. 1993. *The Great Earthquake Experiment: Risk Communication and Public Action*. Boulder, Colo.: Westview Press.

Mileti, D. S. and J. S. Sorensen. 1990. *Communication of Emergency Public Warnings: A Social Science Perspective and State-of-the-Art Assessment.* Oak Ridge, Tenn.: Oak Ridge National Laboratory.

Mitchell, J. K. 1995. Natural Hazards and Sustainable Development. Abstract prepared for the Natural Hazards Research and Applications Workshop, Boulder, Colo.

Mitchell, J. K. 1990. "Human Dimensions of Environmental Hazards: Complexity, Disparity, and Search for Guidance." Pp. 131-175 in *Nothing to Fear: Risks and Hazards in American Society,* A. Kirby, ed. Tucson: University of Arizona Press.

Mitchell, J. K., ed. 1996. *The Long Road to Recovery: Community Responses to Industrial Disasters*. Tokyo: United Nations Press.

Mitchell, K. 1992. Natural Hazards and Sustainable Development. Paper presented at the Natural Hazards Research and Applications Workshop, Boulder, Colo.

Mittler, E. 1997. *An Assessment of Floodplain Management in Georgia's Flint River Basin.* Program on Environment and Behavior Monograph #59. Boulder: Institute of Behavioral Science, University of Colorado.

Munich Reinsurance. 1995. Topics: Natural Catastrophes. Munich: Munich Reinsurance.

National Center for Atmospheric Research, Information and Outreach Program. 1993. Lightning. Boulder, Colo.: NCAR.

National Climatic Data Center. 1996. Lightning statistics. Information from the World Wide Web at: http://www.ncdc.noaa.gov/

National Climate Center and Emergency Mnagement Institute. 1995. *Hurricane Planning for the Atlantic and the Gulf of Mexico.* Instructor's Guide. Washington, D.C.: Federal Emergency Management Agency.

National Institute for Building Sciences. 1997. *HAZUS Technical Manual: Earthquake Loss Estimation Methodology.* Prepared for the Federal Emergency Management Agency. NIBS #5203. Washington, D.C.: NIBS.

National Oceanic and Atmospheric Administration, National Hurricane Center. 1996. The Costliest United States Hurricanes of this Century (unadjusted). Information obtained from the World Wide Web at: http://www.nhc.noaa.gov

National Research Council. 1983. *Risk Assessment in the Federal Government: Managing the Process.* Washington, D.C.: National Academy Press.

National Research Council 1990. *Snow Avalanche Hazards and Mitigation in the United States.* Washington, D.C.: National Academy Press.

National Research Council 1991. *Mitigating Losses from Land Subsidence in the United States.* Washington, D.C.: National Academy Press, Committee on Ground Failure Hazards Mitigation Research.

National Research Council. 1994. *Facing the Challenge: The US National Report to the IDNDR World Conference on Natural Disaster Reduction.* Washington, D.C.: National Academy Press.

National Research Council. 1995. *Flood Risk Management and the American River Basin.* Washington, D.C.: National Academy Press.

Noji, E. K., ed. 1997. *The Epidemiological Consequences of Disasters.* New York: Oxford University Press.

Palm, R. 1983. *Home Mortgage Lenders, Real Property Appraisers and Earthquake Hazards.* Boulder: University of Colorado Institute of Behavioral Science.

Palm, R. 1990. *Natural Hazards: An Integrative Framework for Research and Planning.* Baltimore: Johns Hopkins University Press.

Parker, D. 1995. Disaster Vulnerability of Megacities: An Expanding Problem that Requires Rethinking and Innovative Responses. *GeoJournal* 37(3):295-301.

Peacock, W. G., B. H. Morrow, and H. Gladwin, eds. 1997. *Hurricane Andrew and the Reshaping of Miami: Ethnicity, Gender, and the Socio-Political Ecology of Disasters.* Gainesville: University Press of Florida.

Perry, R. W., M. K. Lindell, and M. R. Greene. 1981. *Evacuation Planning in Emergency Management.* Lexington, Mass.: Lexington Books.

Pielke, R. A., Jr. 1996a. *Midwest Flood of 1993: Weather, Climate, and Societal Impacts.* Boulder, Colo.: National Center for Atmospheric Research.

Pielke, R. A., Jr. 1996b. Reframing the U.S. Hurricane Problem. Unpublished manuscript. National Center for Atmospheric Research, Boulder.

Platt, R. H., ed. 1987. *Regional Management of Metropolitan Floodplains, Experience in the United States and Abroad.* Boulder, Colo.: Institute of Behavioral Science.

Pramanik, M. A. H. 1993. *Impacts of Disasters on Environment and Development—International Cooperation.* INCEDE Report 3. International Center for Disaster-Mitigation Engineering. Tokyo.

Presidential Commission. 1979. *The Need for Change: The Legacy of Three Mile Island. Report of the President's Commission on the Accident of Three Mile Island.* Washington, D.C.: U.S. Government Printing Office.

Prince, S. H. 1920. Catastrophe and Social Change: Based upon a Sociological Study of the Halifax Disaster. *Studies in History, Economics, and Public Law* 94:1-152.

Property Claim Services. 1996. *American Insurance Services Group, Inc. News* Jan. 14.

Quarantelli, E. L. 1982. General and Particular Observations on Sheltering and Housing in American Disasters. *Disasters* 6:277-281.

Quarantelli, E. L. 1983. *Delivery of Emergency Medical Services in Disasters: Assumptions and Realities.* New York: Irvington.

Quarantelli, E. L. 1995. What Is a Disaster? *International Journal of Mass Emergencies and Disasters* 13 (3):221-229.

Quarantelli, E. L. 1998. *What Is a Disaster?* New York: Routledge.

Rackers, D. 1996. *Heat Surveillance Summary—1995.* Jefferson City: Missouri Department of Health, Office of Epidemiology.

Riebsame, W., T. Moses, M. Price. 1986. The Social Burden of Weather and Climate Hazards. *Bulletin of the American Meteorological Society* 67 (11):1378-1388.

Riebsame, W., S. Chagnon, and T. Karl. 1990. *Drought and Natural Resources Management in the United States: Impacts and Implications of the 1987–1989 Drought.* Boulder, Colo.: Westview Press.

Rogovin, M., et al. 1980. *Three Mile Island: A Report to the Commissioners and to the Public.* Washington, D.C.: Nuclear Regulatory Commission, Special Inquiry Group.

Rose, A., J. Benavides, S. Chang, P. Szczesniak, and D. Lim. 1997. The Regional Economic Impact of an Earthquake: Direct and Indirect Effects of Electricity Lifeline Disruption. *Journal of Regional Science* 37(3):437-448.

Rosenfield, J. 1996. Cars vs. the Weather. *Weatherwise* 49(5):14-21.

Rubin, C. B. 1995. Physical Reconstruction: Timescale for Reconstruction. In *Wellington After the Quake: The Challenge of Rebuilding Cities.* Christchurch, New Zealand: Centre for Advanced Engineering and the New Zealand Earthquake Commission.

Schiff, A. J., ed. 1995. *Northridge Earthquake-Lifeline Performance and Post-earthquake Response.* New York: ASCE.

Schneider, S. K. 1995. *Flirting with Disaster: Public Management in Crisis Situations.* Armonk, N.Y.: M.E. Sharpe.

Schuster, R. L., and R. W. Fleming. 1986. "Economic Losses and Fatalities due to Landslides." *Bulletin of the Association of Engineering Geologists* XXIII(1):11-28.

Shah, H. C. 1995. Scientific Profiles of the "Big One." *Natural Hazards Observer* XX(3):1–3.

Sheets, R. C. 1990. The National Hurricane Center: Past, Present, and Future. *Weather and Forecasting* 5(2):185-232.

Simile, C. 1995. *Disaster Settings and Mobilization For Contentious Collective Action: Case Studies of Hurricane Hugo and the Loma Prieta Earthquake*. Doctoral dissertation. Department of Sociology and Criminal Justice, University of Delaware, Newark.

Simon, H. A. 1957. *Administrative Behavior,* 2d ed.. New York: Macmillan.

Sorensen, J. 1991. When Shall We Leave: Factors Affecting the Timing of Evacuation Departures. *International Journal of Mass Emergencies and Disasters* 9(2):153-165.

Sorensen, J., and D. Mileti. 1989. Warning and Evacuation: Answering Some Basic Questions. *Industrial Crisis Quarterly* 2 (3-4):195-210.

Sorensen, J., and G. Rogers. 1988. Community Preparedness for Chemical Emergencies: A Survey of U.S. Communities. *Industrial Crisis Quarterly* 2(2):89-108.

Spangle Associates with Robert Olson Associates, Inc. 1997. *The Recovery and Reconstruction Plan of the City of Los Angeles: Evaluation of Its Use After the Northridge Earthquake.* Portola Valley, Calif.: Spangle Associates.

Stallings, R. A. 1995. *Promoting Risk: Constructing the Earthquake Threat.* New York: Aldine de Gruyter.

Stallings, R. A. 1997. Sociological Theories and Disaster Studies. Preliminary Paper #249. Newark: University of Delaware, Disaster Research Center.

Starr, C. 1969. Social Benefits vs. Technological Risk. *Science* 165:1232-1238.

Storm Prediction Center. 1996. Hail statistics from World Wide Web at: http://www.nssl.uoknor.edu/~spc/.

Task Force on Federal Flood Control Policy. 1996. House Document 465: A Unified National program for managing Flood Losses. 89th Congress. 2nd session. Committee on Public Works. Washington, D.C.: U.S. Government Printing Office.

Topping, K. C. 1998. Model Recovery and Reconstruction Ordinance. In *Pre-event Planning for Post-disaster Recovery,* J. Schwab et al., eds. Planners Advisory Service Report prepared for the Federal Emergency Management Agency. Chicago: American Planning Association.

Trenberth, K. E. and C. J. Guillemot. 1996. Physical Processes Involved in the 1988 Drought and 1993 Floods in North America. *Journal of Climate* 9:1288–1298.

U.S. Department of Agriculture. 1990. *Environmental Data Report.* Washington, D.C.: USDA.

U.S. Department of Commerce. 1995. *A Summary of Natural Hazard Fatalities for 1994 in the United States.* Silver Spring, Md.: National Weather Service, Office of Meteorology.

U.S. Presidential and Congressional Commission on Risk Assessment and Risk Management. 1997. *Framework for Environmental Health Risk Management.* Final report. Washington, D.C.: The Commission.

United Nations Department of Humanitarian Affairs. 1995. Chernobyl: No Visible End to the Menace. *DHA News* (16):2-27.

Watts, M. J. 1983. *Silent Violence: Food, Famine and Peasantry in Northern Nigeria.* Berkeley: University of California Press.

Wenger, D. E., L. Quarantelli, and R. R. Dynes. 1986. *Disaster Analysis: Emergency Management Offices and Arrangements.* DRC Final Project Report No. 34. Newark: Disaster Research Center, University of Delaware.

Wenger, D. E., E. L. Quarantelli, and R. R. Dynes. 1989. *Disaster Analysis: Police and Fire Departments.* DRC Final Project Report No. 37. Newark: Disaster Research Center, University of Delaware.

Wescoat, J. L. 1992. Common Themes in the Work of Gilbert White and John Dewey: A Pragmatic Appraisal. *Annals of the Association of American Geographers* 82(4):587-607.

White, G. F. 1945. *Human Adjustment to Floods.* Research Paper No. 29. Chicago: University of Chicago Department of Geography.

White, G. F. 1973. Natural Hazards Research. Pp. 193-216 in *Directions in Geography,* R.J. Chorley, ed. London: Metheun & Co. Ltd.

White, G. F. 1974. Natural Hazards Research: Concepts, Methods, and Policy Implications. Pp. 3-16 in *Natural Hazards: Local, National, Global,* G. F. White, ed. New York: Oxford University Press.

White, G., and E. Haas. 1975. *Assessment of Research on Natural Hazards.* Cambridge, Mass.: MIT Press.

White, G. F., W. Calef, J. Hudson, H. Mayer, J. Sheaffer, and D. Volk. 1958. *Changes in Urban Occupancy of Flood Plains in the United States.* Department of Geography Research Paper #57. Chicago: University of Chicago Press.

White, G. F., R. H. Platt, and T. O'Riordan. 1997. Classics in Human Geography Revisited: Commentary on Human Adjustment to Floods. *Progress in Human Geography* 21:423-429.

White, R., and D. A. Etkin. 1997. Climate Change, Extreme Events, and the Canadian Insurance Industry. *Journal of Natural Hazards* 16(203):135-163.

Whyte, A. 1986. From Hazards Perception to Human Ecology. Pp. 240-271 in *Themes from the Work of Gilbert F. White*, R. Kates and I. Burton, eds. Chicago: University of Chicago Press.

Wiggins, J. 1996. Letter to the Editor. *Natural Hazards Observer* XX(4):5.

Wilhite, D. A., ed. 1993. *Drought Assessment, Management and Planning: Theory and Case Studies.* Dordrecht, Holland: Kluwer Academic Publishers.

Williams, K. 1996. An Overview of Avalanche Forecasting in North America. Paper presented to the International Avalanche Conference, Davos, Switzerland.

Wisner, B., et al. 1976. Poverty and Disaster. *New Society* 37:546-548.

Working Group on California Earthquake Probabilities. 1990. Probabilities of Large Earthquakes in the San Francisco Bay Region, California. U.S. Geological Survey Circular 1053. Washington, D.C.: U.S. Government Printing Office.

Working Group on California Earthquake Probabilities. 1995. Seismic Hazards in Southern California: Probable Earthquakes, 1994-2024. *Bulletin of the Seismological Society of America* 2(85):379-439.

World Commission on Environment and Development. 1987. *Our Common Future.* New York: Oxford University Press.

World Resources Institute. 1992. *World Resources, 1992-93.* New York: Oxford University Press.

Wright, T. L., and T. C. Pierson. 1992. *Notable Eruptions of Volcanoes in the United States in the 20th Century.* USGS Circular 1073. Reston, Va.: U.S. Geological Survey.

Yezer, A. M., and C. B. Rubin. 1987. *The Local Economic Effects of Natural Disasters.* WP #61. Boulder, Colo.: Institute of Behavioral Science, University of Colorado.

APPENDIX A

Recommendations for Further Traditional Research

CHAPTER 9 PRESENTED A LIST of key research topics that directly address the emergence and implementation of sustainable hazards mitigation. This appendix lists research needs in more traditional hazards and disaster knowledge areas. The two research catalogs are not mutually exclusive, nor are they inconsistent. Sustainable hazards mitigation requires its own base of knowledge, but it also will need advances in most areas of traditional knowledge, packaged for use in innovative ways.

Research needs identified in this appendix are cataloged in the areas of land use; engineering; prediction, forecast, and warning; preparedness; response, recovery, and reconstruction; insurance; economics; and adoption and implementation. Judgments about appropriate research methods to use in conducting a particular piece of research are not offered, so potential researchers can select their own mode of inquiry. Research needs, however, are ranked as being of high, medium, or low priority.

LAND USE RESEARCH NEEDS

Research on land use is needed in the areas of planning, vulnerability assessments, and implementation.

Planning

Research is needed on how to craft land-use plans in a way that will foster and sustain commitment to sustainability. Research is needed on seven topics:

Consensus building and other ways to incorporate the participation of all stakeholders into plan making, adoption, and implementation (high priority).

Mechanisms and processes for building the capacity of local and state governments and nongovernmental organizations to analyze hazards and formulate cost-effective strategies (high priority).

Determination of the factors that will lead to appropriate choices between integrating hazards mitigation into comprehensive plans versus making stand-alone hazards mitigation plans (high priority).

Discovery and understanding of which factors affect the flexibility, effectiveness, and equity of alternative mitigation strategies (high priority).

Determining the outcomes of local hazards mitigation land-use plans in terms of short- and long-term costs incurred by the public and private sectors, gains in reduced vulnerability, and how planning processes, choices, and principles generate community consensus and action (high priority).

Identification of successful regional approaches to land use planning and hazard mitigation (high priority).

Cataloging and understanding how localities value the creation of high-quality plans and how to impact those values (medium priority).

Vulnerability Assessments

Research is needed on how to translate the technology for assessing vulnerability into software and procedures for easy use by local governments. Research should focus on three topics:

How and why (or why not) innovations in existing and emerging hazard assessment technologies affect local land-use planning (high priority).

How to better validate damage functions (what is needed from the natural and social sciences, engineering, and damage estimation modeling) for all hazards (high priority).

What is needed from the natural and social sciences and engineering to provide localities with vulnerability assessments for alternative scenarios of future loss (high priority).

Implementation

Research is needed on at least the following seven topics in order to gain knowledge that can be used to enhance compliance:

Ways to build local commitment to sustainability that are not coercive or costly: persuasion, incentives, monitoring (high priority).

Successful alternative institutional arrangements for land use management—multistate consortia, state hazard safety advisory boards, watershed planning partnerships, and state-local cooperative programs (high priority).

Comparisons of the benefits and costs of land use strategies to avoid development in hazardous areas versus other strategies for safe development versus a combination (high priority).

Factors that help communities enact and enforce land-use restrictions in hazardous areas (medium priority).

The most effective roles for intermediaries and other entities in translating plans into action (medium priority).

Assessing the effectiveness of a variety of state and local floodplain management programs to determine whether they have had a net impact on the value of floodplain areas and whether they affect the way watersheds perform their natural function (medium priority).

The impact/influence of the 100-year standard in many flood protection programs (medium priority).

ENGINEERING RESEARCH NEEDS

Because engineered approaches to hazards mitigation remain a favored national alternative, there are many ideas about the research that should be pursued. These are grouped into standards and codes, research laboratories and equipment, and hazard-specific engineering research needs.

Standards and Codes

The standards and codes in place for all hazards are in need of re-evaluation, assessment, improvement, and enhanced implementation. Further research needs are as follows:

Understanding the key social factors that determine code acceptance and enforcement (high priority).

Reassessment of the purpose of building codes—safety of the occupants, losses from damage to the structure, contents, disruptions in business and lifestyles (high priority).

The most appropriate ways to test and certify criteria for code officials (high priority).

Ways to improve building code performance provisions, including national and state uniformity (high priority).

Ways to integrate scientific input and the views of stakeholders in setting codes and standards (high priority).

How to develop acceptable voluntary standards through a consensus process by a major organization such as the American Society of Civil Engineers or American National Standards Institute (medium priority).

The extent of adoption and adherence to the International Fire Code Institute's model code and the effectiveness of its provisions in specific settings (medium priority).

Research Laboratories and Equipment

The nation's engineering research laboratories and equipment need upgrading, as follows:

A comprehensive plan to use and upgrade existing earthquake engineering laboratories and personnel; integrate new testing approaches into the research infrastructure; consideration to developing a single national testing facility; sharing facilities with other countries (high priority).

An adequate set of testing facilities for the full array of lifelines vulnerable to earthquake damage (high priority).

A full-scale test facility for research for missiles and impact loads for the design of glazing and wall cladding (high priority).

Understanding of the physics of snow drift formation (low priority).

Floods

Specific knowledge is needed to address the nation's predictably most costly natural disaster types:

Financial and other incentives to encourage property owners to adopt floodproofing measures (high priority).

Water management systems that can handle all stormwater events from minor storms where water quality is important to major floods where protection of public health and minimizing economic damage are critical (high priority).

Improved understanding of hydrostatic and hydrodynamic forces on above-ground and basement walls of buildings (high priority).

Design standards for foundations for elevated structures (high priority).

Evaluation and inventory of effective floodproofing measures (medium priority).

Investigation of techniques, procedures, and incentives for coastal and erosion zone structure relocation (medium priority).

Decision support systems for integrated watershed and stormwater management (medium priority).

Evaluation of the joint effectiveness of storage and treatment devices for stormwater quantity and quality management (medium priority).

Droughts

Research is needed for drought that will also aid in the general management of water supply:

Improved real-time control and management of water use for all purposes, including control systems for reducing demand (high priority).

Improved definitions of risk and reliability in water supply (medium priority).

Wind and Hurricane

Research and technology transfer is needed to improve the resistance of buildings to wind loss, specifically:

Methods for providing adequate training and technology transfer to local building officials (high priority).

Retrofitting methods to increase the wind resistance of the current building stock (high priority).

Better standards for the impact and resistance of roofing materials, windows, and doors (high priority).

More data on roofing performance, including performance tests and development of new materials (high priority).

Better definition of wind loads on a variety of structures (medium priority).

Increased performance of wood-frame and masonry-frame structures, at a lower cost (medium priority).

Landslide, Land Subsidence, and Expansive Soils

Mitigation of the various ground failure hazards often is tied to the forecast, warning, or mitigation of other hazards or conditions. Some significant needs are listed below:

Identification of the links between the various forms of landslides and climatic variability (e.g., intense, frequent, and long-duration precipitation) (high priority).

Two- and three-dimensional models of debris flow routes to forecast volumes, speeds, and probabilities of inundation (high priority).

The susceptibility of soil and rock to ground failure as a result of earthquake-induced ground motion (high priority).

Understanding the causes, mechanics, and risk of volcano-related mass movement, including edifice collapse, pryoclastic flows, and lahars (high priority).

Predicting the time-to-failure, installation of warning systems, and mitigative engineering for giant landslides (high priority).

The effects of wildfires on near-surface soils, resulting in increased slope instability during precipitation (medium priority).

Physico-chemical methods to mitigate the effects of swelling soils (medium priority).

Identification of areas subject to ground subsidence and collapse as a result of underground mining, quarrying, and fluid withdrawal (medium priority).

Snow

Research on snow is needed to provide greater insights into adaptive construction practices, specifically:

A "blizzard" map (i.e., joint occurrence of snow and wind) to properly characterize roof snow drift (medium priority).

A proper technical approach and policy for mitigating snow hazard for existing construction (low priority).

Detailed ground snow load maps at a larger scale (low priority).

Earthquakes

The recent creation of three national earthquake engineering research centers will do much to improve the status of earthquake engineering research in the United States. Important future research needs follow.

Determine the effects of ground motions on structures and lifelines, especially of those motions experienced close to the earthquake fault rupture in order to achieve the desired performance of existing and new construction (high priority).

Develop design, analysis construction, and quality assurance procedures to lead to improved seismic performance of new structures and lifelines (high priority).

Reliable and cost-effective techniques to improve the safety of existing seismically vulnerable buildings, bridges, and lifelines (high priority).

Practical and effective procedures to rapidly evaluate, stabilize to prevent subsequent collapse, and repair damaged structures following an earthquake (medium priority).

Innovative technologies to protect structures from the consequences of earthquakes, including the use of seismic isolation, supplemental energy dissipation, active control, and high performance materials (high priority).

Methods for increasing the earthquake resistance of urban lifeline systems. This includes understanding system response, the nature and consequences of interruption of service, and the effects of the interactions of different lifeline systems and facilities with each other. In addition, effective methods for system management, repair, and restoration need to be developed (high priority).

Methods to reduce the vulnerability of urban areas to fire and environmental emergencies instigated by earthquakes (medium priority).

Understand instances where the observed behavior of structures differed dramatically from what was expected based on past earthquake performance and laboratory testing in reference to steel structures and buildings—e.g., steel connections and behavior of steel frame building systems, bridges and buried lifeline elements, evaluation and rehabilitation of existing structures, reconstruction and recovery studies including business and industry restoration, and innovative techniques to control and suppress the spread of urban fires and hazardous materials releases caused by earthquakes (high priority).

Develop methods to mitigate damage to nonstructural elements of buildings (medium priority).

Enhanced understanding of the application of earthquake knowledge (high priority).

Upgrade, modernize, and share facilities and equipment with other nations at experimental earthquake engineering research facilities and accelerate experimental research (high priority).

Develop and implement performance based seismic design—e.g., collapse prevention, life safety, damage control, maintained operations—to allow building owners to define performance levels to meet the specific requirements for the building and its contents (high priority).

PREDICTION, FORECAST, AND WARNING RESEARCH NEEDS

Public alert systems could probably be improved if the process of diffusing existing technology and knowledge were better understood.

Climate Change

Research should be performed to determine how climate change will impact the hazardousness of the nation. The following are needed:

A reference baseline for climate model validation for trends in frequency and severity of weather-related hazards, climate variability, cycles, and warming (high priority).

Predictions of the frequency and intensity of weather-related hazards under a changed climate, using climate model-based studies and paleo/proxy methods (high priority).

Drought

Research is needed to better understand drought's occurrence and to foster social adaptation, specifically:

A comprehensive, consistent drought loss database (high priority).

A better understanding of the social effects of drought; how drought forecasts would be used by decisionmakers (high priority).

Improved forecasting and prediction computer models, especially for atmospheric/oceanic interactions and monitoring (high priority).

A better understanding of the frequency and range of drought at regional and local levels (high priority).

Principles, criteria, and methods for drought control and mitigation plans (medium priority).

Earthquakes

Predicting the times of individual earthquakes does not appear possible in the foreseeable future, so an effective approach to hazard reduction must focus on long-term earthquake potential. Scientists need to address the issues below:

Basic data on faults and earthquakes to define earthquake potential (high priority).

Geological investigations and seismic imaging for fault mapping, trenching of faults to determine dates and sizes of prehistoric earthquakes, seismic recordings and strain rate measurements with global positioning systems and other geodetic techniques (high priority).

An effective strategy, including international cooperation, to test the most important hypotheses about earthquake potential by monitoring and studying quakes everywhere on the planet (high priority).

Identification of the characteristics of a given site that contribute to damage, such as soil conditions (medium priority).

A more complete description of ground motion and its relationship to building response, including new data on shaking and building response, computer models of shaking, and seismograms at sites where damage has occurred (medium priority).

Wildfires

Response planning would be enhanced if wildfire spread could be better predicted. Related needs are listed below:

A stochastic answer to how fires spread (high priority).

Determinations of the flammability of various materials, particularly vegetation (high priority).

Revision and updating of the fire behavior model, incorporating the urban interface (high priority).

Floods

Human response would likely be improved with a better understanding of how to use flood warnings. The following are needed:

Systematic evaluations of warnings, their specific impacts, and their usefulness (high priority).

Better understanding of the connection between awareness and action, with a focus on public awareness programs, graphics, and other communications techniques (medium priority).

Better understanding of how long- and short-range flood forecasts are used for local policymaking (medium priority).

Hurricanes

The following pointed research needs exist regarding the prediction, forecast, and warning for hurricanes:

A denser network of observing stations along the Gulf and Atlantic coasts to detect the strongest winds (high priority).

Improved knowledge of response in moderate- and low-risk locations, tourist response, vertical refuge and early evacuation, predicting evacuation destinations, and predicting the quantitative effect of a hurricane on a community (high priority).

Improved remote observation platforms: better algorithms for Doppler weather radar for determining two-dimensional wind fields over land and better equipment for aircraft and satellites for measuring surface winds over water (high priority).

Better understanding of the reliability of SLOSH, with attention to wave action (high priority).

Quantification of the frequency of various conditions on roadways during evacuation; increased collection of real-time traffic count data in evacuations (medium priority).

A comparison of the effectiveness of various modes of notification, including automated telephone recordings, door-to-door, self-activating radios, and cable television (medium priority).

The most effective means of educating the public about a hurricane threat and evacuation (medium priority).

Improved forecasts of forward speed and intensity (medium priority).

Empirically based rationale for per-person space allotments in temporary shelters (low priority).

Technological Hazards

The causal links between natural and technological hazards are not well understood; several needs are evident:

Automated decision support systems, integrating physical and behavioral models with real-time instrumentation to predict exposure (high priority).

Better understanding of the ways in which natural disasters induce technological hazards, especially engineering investigations of releases to better understand the causal mechanisms and establish a database of incidents for modeling (high priority).

Better understanding of compliance with shelter orders (high priority).

An update of the Environmental Protection Agency's 1988 study of community preparedness for chemical emergencies, including warning systems (medium priority).

Tornadoes

Research in the following areas is needed to reduce the number of deaths and injuries from tornadoes:

Ways to use new technologies and datasets, such as that from WSR88-D, GOES-8 and -9, and wind profilers, to improve prediction (high priority).

Refinement of high-resolution, real-time thunderstorm models to predict the imminent formation of tornadoes (one to six hours) (high priority).

New, combined algorithms for forecast and detection, by blending WSR88-D data with other observations (high priority).

Multidisciplinary research for practical mitigation, tornado-resistant construction, and in-residence shelters (high priority).

The critical warning needs of local officials, the news media, and the public (medium priority).

Information about the telecommunications and computer capabilities of local emergency management officials (high priority).

Documentation of the relationships among tornado characteristics, warning lead time, appropriate public responses, and morbidity and mortality (high priority).

Assessment of the warning services incorporated into National Weather Service operations (high priority).

An inventory of the social benefits of new forecast and warning products and services (medium priority).

Tsunamis

Several key research targets exist for the infrequently occurring tsunami hazard:

Assessments of the nature and extent of the tsunami hazard at selected localities (high priority).

A better model to describe the characteristics of tsunamis formed as a result of thrusting and landslides, including improved modeling, tectonic studies, and data on tsunami travel times (medium priority).

Techniques for estimating expected amplitude along with estimated time of arrival of a tsunami (medium priority).

Deep-ocean gauges for data on wave heights from the open ocean and broadband multipurpose sea-level detectors in the open coast and deep sea as well as harbors (low priority).

Volcanoes

Improvements in the prediction, forecast, and warning for volcanic eruptions could result from research into the following areas:

Improvement of real-time computer analysis to predict eruptions (high priority).

Improvements in seismic technology (high priority).

Continued study of eruption and emplacement mechanisms (high priority).

Automated recognition of volcanic/hydrological events and improved modeling of volcanic ash dispersal (high priority).

Refinement of the interpretation of seismic signatures as indicators of eruption (high priority).

Use of satellite radar to monitor ground deformation (high priority).

Enhanced understanding of activity at large restless volcanoes (high priority).

Refinements in deformation source models (medium priority).

Information about magma-groundwater interactions (medium priority).

Real-time global positioning system networks (medium priority).

Detailed studies of infrastructure impacts (medium priority).

Improved communication among scientists, authorities, and emergency planners for warnings and public response (medium priority).

Social Response

Many research questions cut across many or all hazards, especially those that have to do with human activity in the face of warnings. Several specialized topics deserve research attention:

The most effective way to inform the news media of the factors important in presenting emergency and warning information to the public (high priority).

How uncertainties detract from sound decisionmaking and what aids would help (high priority).

An integrated approach to warnings systems research, including state-of-the-art knowledge of human response and full analysis across hazards (high priority).

More information about the unique needs for public warnings of fast-moving events such as flash floods and chemical spills (high priority).

Incentives and constraints to local adoption of warning systems (high priority).

Whether the Emergency Alert System improves warning effectiveness (high priority).

The potential of a new theoretical framework for warning dissemination, including information reception, understanding, response by officials, information sharing, two-way communication (medium priority).

How, to what extent, and what kind of preemergency public education affects people's response to warnings (low priority).

The influence on public response privately issued forecasts and warnings for weather related events (high priority).

RESEARCH NEEDS FOR PREPAREDNESS, RESPONSE, RECOVERY, AND RECONSTRUCTION

Preparedness

Although much has been learned about emergency preparedness, research is needed to produce new knowledge on the following topics:

How states plan for and respond to disasters and the extent to which federal-state agreements have fostered improvements (high priority).

How to improve research dissemination to practitioners (high priority).

Planning, training, skills, and maintenance needs at the local level (high priority).

Ways to increase community support for disaster preparedness (medium priority).

Which preparedness activities are undertaken by manufacturing and service organizations, hospitals, schools, utilities, and others (low priority).

Whether particular organizational strategies foster more comprehensive preparedness (low priority).

Response

There are still stubborn puzzles about both individual and organizational response behaviors. And changing cultural and technological environments bring about new issues to tackle, such as:

How cultural and social inequality, diversity, politics, economics, and technology affect preparedness and response (high priority).

The sheltering of institutionalized populations, communitywide coordination of housing, the appropriateness of mobile home parks as postdisaster housing, the psychological consequences of damaged or destroyed housing on victims (high priority).

How new technologies such as geographic information systems and electronic communications are disseminated, used, and affect response (high priority).

Risk factors for injury and postdisaster illness (medium priority).

Cost-effective search and rescue approaches, structures, and techniques (medium priority).

Mass communication in disasters, including changes in the gatekeeping function of the news media; effective communications channels; differences between local and national disaster reporting; impacts of changes in news media technology (medium priority).

Cross-national comparative research (medium priority).

Clarified criteria for judging organizational and system performance in disaster emergency medical services (medium priority).

How sheltering and housing are experienced and undertaken at individual, group, organizational, and community levels and how outside groups affect that process (medium priority).

Recovery and Reconstruction

Recovery is one of the least well understood aspects of the hazards field. The more pressing needs for advancement in this area are as follows:

A determination of the factors that contribute to effective, efficient recovery and those that constrain it (high priority).

The needs of groups that suffer from lingering and/or late-blooming effects of disasters (high priority).

How local governments can effectively plan and manage recovery and reconstruction, particularly in terms of mitigation planning (high priority).

More attention to decisionmaking at all levels, testing of sustainability as a sensitizing model for recovery, monitoring and evaluating progress in sustainable development during recovery (high priority).

Techniques for providing communities with pre- and postdisaster information and expertise to recover effectively (medium priority).

Investigation of why the practice of recovery and reconstruction is so poor, given the knowledge, written products, and training courses available (medium priority).

RESEARCH NEEDS FOR INSURANCE

Several key questions remain about both the specifics of using insurance to spread (or mitigate) losses from hazards and about the secondary effects that widespread reliance on insurance would have on the nation's insurance and financial sectors.

Obtaining Additional Information

The following types of information would help define the role of insurance in mitigation:

Specification of cost-effective mitigation measures that could be applied to new and existing structures; incentives that insurers can offer for them (high priority).

Improved modeling of events to guide risk selection, pricing, financing (high priority).

Reliable measures for insurers and public decisionmakers of the cost effectiveness of mitigation efforts (high priority).

Simulations of the impact of different disasters on insurers' profitability, solvency, and performance (medium priority).

Organizational Arrangements

Several puzzles exist about the organizational arrangements needed for insurance to have an appropriate hazards mitigation role:

The best way to measure insurance industry capacity and the impact of catastrophes on the ability of insurers and reinsurers to generate new capital (high priority).

The appropriate tradeoffs among mitigation, insurance availability, and insurance affordability to make mitigation a focus of state insurance regulation (medium priority).

Whether mitigation will ever be sufficient to make normal insurance markets feasible in the highest-risk earthquake and hurricane zones (medium priority)

The roles of insurance, disaster relief, and other programs in recovery; how uninsured people and businesses finance disaster losses (medium priority).

RESEARCH NEEDS OF ECONOMICS

Next to saving lives, the most compelling and frequently cited reason for investigating hazards and disasters is to reduce their costs. Ironically, the economic aspects of hazards, disasters, mitigation, insurance, and related issues are perhaps the least well understood. The need for thorough knowledge of the financial implications of hazards—at both micro and macro scales—is profound. Some of the more pressing needs are listed below:

Better estimates of the economic benefits and costs of all mitigation alternatives (high priority).

Better estimates of nonmarket losses such as personal memorabilia, cultural assets, historical monuments, and natural resources (high priority).

Far more precise damage functions (high priority).

Better estimates of the costs to a regional economy of loss of public facilities and infrastructure (high priority).

Assessment of the full costs and benefits (including environmental ones) of occupying hazard-prone areas (e.g., floodprone lands; high priority).

Better estimates of how firms respond to disaster-caused loss of markets and/or critical suppliers (high priority).

Incentives that would encourage the development and use of loss information (high priority).

The implications for federal insurance-related policies (and the supply of insurance) if future financial centers become highly diversified and integrate investments, banking, and insurance (medium priority).

Better estimates of how insurance affects real estate markets (medium priority).

Better estimates of how household finances are impacted by disasters (medium priority).

RESEARCH NEEDS FOR ADOPTION AND IMPLEMENTATION

There is still a need for both a more comprehensive theoretical formulation and research to clarify and extend present knowledge about mitigation adoption and implementation.

Impact of Perceptions

We still do not fully understand the link between perception and action. The following is needed:

Identification of the variables that connect risk perception with adoption of mitigation: additional perceived characteristics of natural hazards, heuristics, and demographic variables (medium priority).

Processes

More knowledge is needed about the human processes that underlie taking steps to reduce future disaster losses, specifically:

Better understanding of the ways in which contextual factors affect mitigation behavior; and the role of emotion, cognitive processes, information, social processes (high priority).

More and better-designed evaluations of persuasive awareness programs (medium priority).

Identification of the salient characteristics of alternative mitigation activities to help predict adoption and implementation (high priority).

Field experiments to examine the effects of persuasive appeals on mitigation in situations of low threat, with a longitudinal component (high priority).

Principles for communicating technical and scientific information about hazards to policymakers, followed by empirical testing and field evaluation of their effectiveness (medium priority).

The degree to which models of mitigation behavior can be generalized to racial and ethnic minorities in the United States and to other cultures (medium priority).

Adoption and implementation processes in organizations: evaluation of alternatives, perceptions of hazard risks at facilities, willingness to allocate resources (medium priority).

Approaches

Additional knowledge is needed to better inform attempts to further the loss reduction actions of others, specifically:

The relative influence of voluntary and involuntary approaches to hazard mitigation and how to select or combine the appropriate ones (high priority).

The extent to which legal mechanisms are actually used to influence mitigation (medium priority).

A better understanding of knowledge transfer processes among hazards researchers and practitioners, along with mechanisms for that transfer: types of information most needed by different categories of practitioners and researchers; and the most effective ways of packaging research information (medium priority).

APPENDIX B

Impacts of the 1975 Assessment

THE FIRST NATIONAL ASSESSMENT of natural hazards (White and Haas, 1975) described the natural hazards research in the early 1970s as spotty, uncoordinated, and dominated by physical and technological fields. It also observed that the United States had done relatively little about the economic, social, and political aspects of adjustment to natural hazards and that the hazards-related social research that had been done was only occasionally applied to the activities of entities involved in disaster prevention or recovery. White and Haas also critiqued much of the social research as sporadic, noting that it was limited to the interests of individual investigators or the missions of specific agencies and sometimes carried out by agency staff members who were not necessarily qualified researchers. Occasionally, they observed, research was "farmed out" to academic groups or research and development firms but with mixed results. Furthermore, it was noted that state, local, and voluntary agencies had no money to sponsor research and that no agency provided basic funding or coordination for natural hazards research in the nation. No broad body of knowledge had been created, nor had earlier research findings been updated to account for social and economic changes.

Looking back on these observations and criticisms, it can be seen that the 1975 assessment had considerable impact on the hazards community and on the nation over the past 20 years. But much of the natural hazards research that has been done since 1975 reflects changes in societal, political, legislative, economic, juridical, and scientific attitudes and realities—shifts that the staff of the 1975 assessment project could not have anticipated. In effect, rapidly moving contemporary history overtook many of the findings and recommendations published in 1975. And it is difficult if not impossible to ascribe many of the subsequent developments to the 1975 assessment itself.

Some impacts of the 1975 assessment are speculated on below. The 1975 assessment should be viewed not as a success or a failure but as what it was: a guidepost for those who have research resources for natural hazards and for those who are the ultimate users and beneficiaries of the research findings and recommendations.

IMPACTS ON RESEARCH, POLICY, AND MANAGEMENT

The 1975 assessment did affect the hazards and disaster research that has been done since then, the government policy that has emerged, and actual mitigation and disaster response operations throughout the nation. The 1975 assessment had a strong and positive impact on the field of hazards and disaster research, although not necessarily on its dissemination and implementation. This impact is the result of two factors. The first is the continued involvement and leadership in the field on the part of the graduate students who participated in the 1975 assessment. The second is the ongoing participation of academicians and practitioners at the Natural Hazards Research and Applications Information Center's (NHRAIC) annual workshops. If there is any ongoing overarching approach to bringing disaster and hazards research into common focus, it is the deliberations at that and other more hazard-specific conferences and through its publications. NHRAIC's clearinghouse function has also encouraged federal agencies to fund a small amount of disaster research and has generated an interest on the part of academicians and planning or research firms in doing such research. Researchers doing natural hazards-related research of the kind recommended by the 1975 assessment have been affiliated with many universities and private entities throughout the nation.

It can be argued, however, that over the past two decades the focus of performed research has been steered less by recommendations than by the development of new forecast technologies and information systems and by a concern for technological and environmental hazards that largely transcended and co-opted concern for natural hazards, except when a truly spectacular natural disaster occurred.

To ask if the original assessment had any impact on disaster policy produces three answers: "yes," "no," and "wait and see." The 1975 assessment report pointed to five major methods of adjusting to natural hazards common to most hazards—relief and rehabilitation, insurance, warning systems, technological aids like protective works, and land-use management. In hindsight it seems odd that the term "hazards mitigation" was not used since it has become the battle cry of the late 1990s. The first assessment led to an important change in national hazards policy insofar as mitigation is concerned, but it cannot be said to have thoroughly permeated the state and local levels.

The wait-and-see response comes because hazards mitigation may now be federal policy and may be gaining favor in some states, but the converse is often also true. After Hurricane Hugo, South Carolina's legislature refused to provide funds for posthurricane mitigation planning (as required by federal law). In other states there are thousands of homes, condominiums, resort businesses, and other development built directly in the unprotected path of the next hurricane.

It is difficult to evaluate the impact of natural hazards research on state disaster plans over the past two decades. It took a number of years—much longer than the two years prescribed in the 1975 assessment's recommendation—for some states to develop such plans and, when they did, implementation was a sometime thing, hindered by inadequate staffs and small budgets. Few of the plans went beyond emergency preparedness and response to include recovery or mitigation.

Where research does seem to have had an impact is through the efforts of such groups as the Association of State Floodplain Managers and where federal agencies such as the National Oceanic and Atmospheric Administration, Federal Emergency Management Agency, U.S. Army Corps of Engineers, and U.S. Geological Survey have tried to combine technical progress and new knowledge with the findings of social research in their dealings with state and local planners. A good example of this is the way hurricane evacuation planning utilizes not only technical mapping but also information derived from the Florida State University studies of evacuation planning and response.

It would be worthwhile to compare state and local utilization of research done by physical scientists and engineers with that done by social scientists. The 1975 assessment included both approaches in its recommendations, and, as a result of activities associated with NHRAIC's clearinghouse, the linkage is much more profound now than it was in the early 1970s. In fact, some of the linkages proposed in 1975 are now actually required by law!

Although the original assessment devoted considerable attention to warnings, emergency response, and relief and rehabilitation, its impact on operational practices is not easy to evaluate. Research has been slow to have an impact on practice and operations, and little of the impact that has occurred can be traced to the 1975 assessment per se but to the kinds of research it engendered. Such impacts can be found in evacuation planning, hazards assessment methodology, warning systems, and operational linkages between emergency managers, urban planners, state attorneys general, and other public officials with whom they traditionally had had little in the way of a working relationship.

CONSTRUCTING THE "HAZARDS COMMUNITY"

The 1975 assessment had a profound impact by creating a "hazards community." Before 1975 the hazards field was fragmented and consisted largely of the Federal Emergency Management Agency's predecessors, the American Red Cross and other voluntary response groups, the National Weather Service, the U.S. Geological Survey, the U.S. Coast Guard, and the military. At the state and local levels the emphasis was on civil defense, although there was some "dual use" of plans, facilities, and communications systems.

Into this milieu came the 1975 assessment and its offspring, the Natural Hazards Information Center. Of all NHRAIC's accomplishments, the most significant and far reaching has been its years of cross-pollination—the involvement of people from a variety of fields of expertise and governmental responsibility, researchers, and practitioners—to create a sense of community in the hazards and disaster field. In addition to its ongoing clearinghouse services and annual workshops, NHRAIC has facilitated or conducted planning and discussion sessions for individual groups and for those with related narrower concerns.

SETTING THE NATION ON A PATH

The 1975 assessment provided a positive but perhaps naive perspective: there are many things we can and need to do to reduce the natural hazards threat to our nation, so here's a list of research that needs to be funded and let's get with it. That perspective is still with us, although it is tempered somewhat by 20 years of experience laced with successes and frustrations.

Limited progress notwithstanding, the first assessment did give the hazards community a healthy perspective—that we need not devote all of our efforts to picking up pieces after a disaster because we can ameliorate the impact of catastrophic events by using better planning, land-use controls, and other preventive and mitigation measures. Hazards mitigation is the first order of business and the cornerstone of the nation's approach to hazards today.

The 1975 assessment did not succeed in raising the level of funding for hazards research, especially in the social sciences.

Impacts of the Proposed Research Agenda

White and Haas's (1975) assessment outlined a 10-year research agenda that included five major research strategies, a proposal for cross-hazards research into issues common to most hazards, a number of subsets within those, and a list of 152 specific physical and social science research opportunities related to many natural hazards.

Postdisaster Audits

It was recommended that a number of predesignated interdisciplinary teams be readied to do postdisaster audits in perhaps 10 affected communities. The audits would examine each area's predisaster state of preparedness, any special circumstances involved, the emergency response, and the recovery process. The postaudits were to provide a compendium of information about communities in disaster that could be used to identify differences, similarities, and special conditions that needed further study.

This recommendation was not carried out, at least not within the parameters prescribed in the 1975 assessment. Most of the postaudit or similar studies that have been done have involved four or five communities. Only two studies actually undertook larger sampling, and they were narrowly focused (one on warning systems, one on recovery). Many "partial" postaudits have been performed, however, and their combined information can be useful to emergency managers and community planners. A compendium of all those findings could well serve the purpose of postaudits as envisioned by White and Haas, especially if it were to offer not only the theoretical lessons learned but also some practical applications.

It can be said, therefore, that the 1975 recommendation for comprehensive postaudits did, in fact, encourage various researchers and federal and relief agencies to undertake a variety of limited but informative follow-up evaluations. But the theoretical and practical lessons that could be gleaned from all of the partial postaudits done since 1975 have not been fully cataloged.

Longitudinal Studies

There have been no comprehensive longitudinal studies of disasters since the first assessment called for them. There are, however, anecdotal longitudinal-type lessons to be found in a few visionary and prescriptive documents written by experts, and in "anniversary of the disaster" and other feature stories in the popular news media.

Clearinghouse Service

Perhaps the most understated but ultimately productive recommendation of the 1975 assessment was the establishment of a national natural hazards information clearinghouse. The original reason for proposing a clearinghouse was to ensure the rapid and wide circulation of information and judgments among the producers and users of natural hazards research and to provide current research findings to local and state planners and to the postdisaster "recovery councils" called for under the Disaster Relief Act of 1974. This recommendation was the genesis of NHRAIC, which has functioned effectively as a source of information on natural hazards history, disaster planning and response, mitigation, and research findings and since 1976 has played a role that goes far beyond information exchange.

A Strategy for the States

This recommendation was based on 1974 federal legislation that provided each state with funds to develop disaster preparedness and prevention plans and directed the establishment of statewide disaster recovery councils. The 1975 assessment recommended that state agencies be targeted for receipt of available research findings pertinent to that state's disaster problems, along with ideas about how state plans could coordinate with those of federal agencies, voluntary groups, and neighboring states. Although the projected state recovery councils were never funded and were dropped from subsequent federal disaster legislation, there is no gainsaying NHRAIC's value as a clearinghouse for state planners.

Congressional Overview

This recommendation recognized the balkanized nature of congressional oversight of the many ways in which federal, state, and local government agencies deal with natural hazards, how they relate to one another in so doing, and the role of the insurance industry and other regulated parts of the private sector. White and Haas acknowledged that it would be impractical to combine all of these activities into a single agency responsible to one committee in each house of Congress. But they called for a "blanket review" by one or more committees that would examine all of the activities, how they relate, what seems to be neglected, and what needs closer integration. This, they thought, would help coordinate the efforts and assure that the sparse federal funds available for natural hazards research came close to achieving maximum net returns to the nation.

It never happened. Congress did approve President Carter's initiative to combine a number of agency components and programs into FEMA, but this has not resulted in any discernibly coordinated approach to congressional oversight of disaster research needs. The Congress has, in fact, done little to encourage agency funding of disaster research except in relation to earthquakes and most recently has reduced budgets for or proposed outright elimination of some major areas of disaster-related physical science studies. Congressional action actually tends to skew hazards research agendas toward an issue that is the subject of current public and media focus, rather than unify them.

Impacts of Specific Research Recommendations

Almost 150 research recommendations were listed in the 1975 national assessment. The impacts of the first assessment regarding event notification were quite modest, with a trend toward paying more attention to nonstructural flood mitigation. What was seen as promising research on hurricane modification was abandoned. Many of the flood research opportunities were related to the technology of warning, mapping, and protective devices, and considerable work has been done in those areas since 1975.

Every single area of the 1975 recommended research on earthquakes has been the subject of considerable research over the past two decades. Whether the findings of such research have been implemented would take considerable probing because of the factors involved in changing building technology and codes.

There has been continuing research into severe storm warnings, hurricane forecasting, and earthquake prediction and revised probabilities. Research over the past two decades in the social and behavioral sciences on the human dimensions of warning systems and warning response has blossomed. In fact, we probably know more about public response to risk information and warning than about anything else in hazards research. Moreover, that knowledge has been put to good use and is likely the reason why the number of deaths from disasters in the United States has decreased over the past 20 years.

The original assessment claimed that the most promising short-term research thrust was testing and refining new approaches to land-use management by studying ways to speed up its adoption and measure its social effectiveness. Over the past two decades there has been some scattered research on this topic, but it still needs much attention.

The whole question of insurance as part of the hazard mitigation and disasters relief mix has changed drastically since 1975. At that time the major question was whether the availability of disaster relief kept people from purchasing flood or earthquake insurance. Since then research has examined a wide range of related issues but has not definitively established whether it does or not. Only now—two decades later—is a new evaluation of insurance being undertaken. The 1975 assessment recommended research into code development and enforcement, which now is becoming a major issue for researchers and the insurance industry.

The 1975 assessment posed several research questions about relief and rehabilitation, including concern for various levels of assistance and their impacts among socioeconomic groups. Soon after 1975 panels to deliberate how disaster victims were assisted, how to deal with mental health problems, and the adequacy of the way help was provided bulked large on the agendas at the annual hazards workshop. A number of researchers were active in this subject area, and many of their findings and recommendations did find their way into Red Cross operational manuals and affected the way federal disaster programs were implemented. But research in this area has waned, and today there is less national discourse about meeting human needs after a disaster than there was two decades ago.

The thrust of the 1975 assessment was that researchers should seek out nonstructural alternatives to structural and control works. These recommendations seem to have gone forward, and research findings have had some impact as witnessed by a preference for relocating at-risk homes and communities after the 1993 Midwest floods.

The advent of the computer, geographic information system, satellite observations, and other innovative technologies has made today's data collection and analysis methods far more sophisticated than anything envisioned in the 1975 assessment. The big issue now is making sure practitioners know how to use the technology and the mass of information becoming available to them.

As the first assessment recommended, the nation has pursued extensive research into building design and performance and code development. The insurance industry in particular has made a major commitment to code enforcement and inspection. The National Science Foundation has funded a great deal of earthquake engineering research.

The 1975 assessment also noted that little is known about why people choose among various possible adjustments and that it is often assumed people will act in a certain way before disasters but that when the real situation arises they act differently. Since the 1975 assessment there have been many studies of pre- and postdisaster behavior, but they have tended to focus either on response to emergency warnings and evacuation or problems with the relief process and relief agencies. Although research has documented that people do make different choices about adjustments, we still do not know why. The question also exists at higher levels of human aggregation, such as business and government.

DISSEMINATION OF FINDINGS

The dissemination of research findings was not emphasized in the 1975 assessment, perhaps because the focus was on future research needs. Today it is a vitally important component of hazards work, and a common plea of practitioners is that the research information needs to be packaged in meaningful, understandable terms. The lack of readily useable published findings notwithstanding, much of the research eventually does get used by practitioners.

EXPANDING RESEARCH CENTERS

The nation's hazards research infrastructure has changed significantly since 1975. First, there are many more research centers in place today than 20 years ago (including most of those that were in operation in 1975), largely because the original centers recruited and supported many graduate students who have remained committed to natural hazards. Many of those students are now among the nation's leading hazards experts, and they both staff the centers that were in place 20 years ago and have started centers of their own. Second, there are fewer graduate students in training today to replace their mentors than was the case when the first assessment was completed in 1975. Fewer young people are choosing careers in the physical sciences today, and funds to support them are more scarce.

Below is a short list of some of the nation's centers for hazards research and mitigation. It is not comprehensive but merely intended to illustrate the centers' diversity.

Center for Advanced Decision Support for Water and Environmental Systems (CADSWES), University of Colorado, Boulder

Center for Earthquake Research and Information (CERI), University of Memphis

Center for Hazards Research (CHR), Department of Geography and Planning, California State University at Chico

Center for Technology, Environment, and Development (CENTED), Clark University

Coastal Engineering Research Center (CERC), U.S. Army Engineer Waterways Experiment Station, Vicksburg, Mississippi

Consortium for International Earth Science Information Network (CIESIN)

Cooperative Institute for Research in Environmental Sciences (CIRES), University of Colorado, Boulder

Disaster Management and Mitigation Group (DMMG), Oak Ridge National Laboratory

Disaster Research Center (DRC), University of Delaware

Earthquake Engineering Research Center (EERC), University of California at Berkeley

Fluid Mechanics and Wind Engineering Program (FMWEP), Colorado State University

Hazards Assessment Laboratory (HAL), Colorado State University

Hazards Research Laboratory (HRL), Department of Geography, University of South Carolina

Hazard Reduction and Recovery Center (HRRC), Texas A&M University

Institute for Crisis and Disaster Management, Research, and Education (ICDMRE), George Washington University

Institute for Disaster Research (IDR), Texas Tech University

International Hurricane Center (IHC), Florida International University

John A. Blume Earthquake Engineering Center, Stanford University

National Center for Earthquake Engineering Research (NCEER)

Jet Propulsion Laboratory (JPL)

National Drought Mitigation Center (NDMC), University of Nebraska-Lincoln

Pacific Emergency Management Center (PEMC), The Simeon Institute

Scripps Institute of Oceanography

Southern California Earthquake Center (SCEC)

Wharton Risk Management and Decision Processes Center (WRMDPC), University of Pennsylvania

Wind Engineering Research Center (WERC), Texas Tech University

APPENDIX C

Putting Knowledge Into Practice

A SURVEY OF RESEARCHERS AND PRACTITIONERS

PAST INVESTIGATIONS HAVE SUGGESTED that research knowledge is most effectively put to use when researchers and practitioners communicate with each other, why they work together throughout the research process, and when researchers produce a subset of products specifically written for users.

As part of the second assessment project, hazards researchers and practitioners were surveyed about the knowledge transfer processes used since the 1975 assessment. A total of 50 researchers and 28 practitioners were interviewed with a pretested schedule administered by trained interviewers. The average interview time was 25 minutes, and most were conducted by telephone.

Respondents were asked a series of questions about the information transfer process. Researchers were asked how they disseminated their own findings, what dissemination methods they thought were the most and least effective, and if their findings were used. Practitioners were asked if they had used research findings, how they came to use them, what they thought the impact of each application had been, and if they had passed the information on to other practitioners. The following sections report on what was found.

Knowledge Transfer Mechanisms Used

The written method of knowledge transfer most frequently reported by researchers was to publish findings in traditional academic journals, monographs, books, chapters, and conference proceedings. They also reported writing for practitioner-accessible publications. The oral dissemination technique cited most often by researchers was making presentations at practitioner conferences and workshops.

Local government practitioners indicated that they most frequently received information directly from dissemination organizations, through personal relationships, and at conferences and meetings. None reported receiving findings from academic journals—the transfer mechanism most frequently used by researchers. Federal- and state-level practitioners also reported that they are increasingly using e-mail and the Internet to obtain and disseminate new information.

Private practitioners rely on the same knowledge transfer mechanisms, but they reported placing greater reliance on their own initiative. Private-sector practitioners also pass information directly to their clients by summarizing and interpreting findings both in documents and direct consultation.

Evaluation of Knowledge Transfer Mechanisms

Researchers thought that conferences, meetings, and workshops were effective ways to disseminate findings to practitioners. They reported that interpersonal contacts were crucial for effective knowledge transfer. Researchers realized that peer-reviewed journal articles are the least effective mechanism for bringing information to practitioners. But they also believed that such journal publication has its role: getting feedback and making improvements before the new information is shared with practitioners, and as a long-term repository of knowledge. Researchers consider government publications and products like educational pamphlets and nontechnical brochures to be highly effective.

Federal and state practitioners perceive networking at conferences and workshops as the most effective means of acquiring new research information. One respondent replied that "interaction leaves a legacy of knowledgeable people and strong advocates for particular mitigation strategies." Face-to-face interactions work best in situations where researchers present their findings and practitioners are encouraged to ask questions.

Use and Impact of Research Findings

Many of the researchers admitted not knowing whether their findings were used. The reasons for this included the difficulty of finding out whether their findings have been used, and the fact that implementation is often a gradual, cumulative process.

Many small-government practitioners reported using research findings to boost the credibility of proposed plans for city development, for obtaining mitigation funds, and to justify the implementation of stricter standards and building codes. Federal and state practitioners credited research findings with helping to improve problem solving, decision-making, products, and public policy. Many documents are periodically revised to include new research findings, including public information releases, response plans, and materials distributed to the community.

Factors that Affect Knowledge Transfer

Researchers and practitioners alike reported four factors that affect knowledge transfer: culture, institutions, links, and interaction.

Culture included such items as language, goals, schedules, styles of communication, and accessibility. Both groups agreed that there are cultural differences between researchers and practitioners—that they live in different worlds; they are both well-meaning and hard working but have different priorities. A predominant cultural issue was language. The remarks about language revolved around the problems of technical jargon and writing that is too academic. Although some researchers did not feel that it was their responsibility to write in accessible language, most acknowledged it was something they would change if they had the time, money, energy, and incentive.

Institutional barriers to knowledge transfer include all of the restraints, rewards, and boundaries that accompany work in the research and practice communities. Practitioners and researchers alike recognized institutional constraints to knowledge transfer—for example, researchers focusing on pure rather than practical research in order to keep their university jobs and get tenure. Structural and institutional barriers also affect practitioners by limiting time, money, staff, and other resources. A federal practitioner summed up: "The reward system on both ends is prohibitive; researchers are not rewarded for getting stuff to practitioners, and practitioners are not rewarded for seeking out new research."

Links include people, projects, workshops, organizations, consultants, education and information centers, and various forums that make research more accessible. Links between the research and practitioner communities were mentioned even more frequently by practitioners than by the researchers. A local practitioner said that there are not enough sources of information transfer—practitioners just do not know where to go to get the information they need.

One observation made by almost all the locals—but not by any of the other practitioners or the researchers—was that information is not reaching the local end user and the problem was in the path from federal to state to local level. The workshops that are effective for knowledge transfer are always held in big cities, never in towns.

Many of the researchers mentioned the importance of and need for more interaction between practitioners and researchers. Some said that each group was largely intimidated by the other. Some claimed that the obstacle may not be in getting the information from researcher to practitioner but in getting the practitioner to implement it once it is received. And two engineers commented that there is no financial incentive to interact with practitioners for the dissemination of new findings.

Many practitioners commented that there is no connection between what practitioners need and what the researchers investigate. Overwhelmingly, the practitioners recommended joint projects, integrated agendas, reaching compromises on priorities, and having more user-oriented research. Many suggested interpersonal relationships as a way to bridge the gaps.

Conclusion

The survey of hazards knowledge transfer confirmed several common perceptions about the process and validated some of the conclusions of previous research on the topic. Existing efforts at knowledge transfer, like workshops, should be continued. But it is clear that more is needed. Knowledge transfer results when the two worlds of research and applications are able to blend their work, but doing so requires extraordinary researchers and extraordinary practitioners who are able to step out of their paradigms and effectively communicate with each other. Many such people exist, but their numbers are still too few.

The nation needs more professionals and mechanisms in the stratum between the research and practitioner communities to achieve sustainable hazards mitigation. More organizations should be dedicated to—and people rewarded for—carrying research findings from researchers to users and carrying the practitioner's viewpoint into the creation of more useful research.

Knowledge Transfer Organizations

Many kinds of organizations are in place now or have been successfully tried over the past two decades to transfer hazards information. Each of the selected knowledge transfer organizations listed below has been successful at accomplishing what it set out to do and represents many others that are not named here. Most important, they illustrate an existing capability for implementing sustainable hazards mitigation.

Association of Bay Area Governments (ABAG). The ABAG is one of more than 560 regional planning agencies across the nation that works to solve problems in areas such as land use, housing, environmental quality, and economic development. Owned and operated by the cities and counties of the San Francisco Bay Area, the ABAG was established in 1961 to protect local control, plan for the future, and promote cooperation on areawide issues.

Board on Natural Disasters (BOND). The BOND coordinates the National Research Council's advice to the federal government and others on disaster reduction. The BOND has one standing committee, the U.S. National Committee for the Decade for Natural Disaster Reduction. It represents the United States in the world's Decade activities. The committee reflects a national commitment to mitigation and disaster reduction and to the integrated efforts of the basic and applied science communities.

Building Seismic Safety Council (BSSC). The BSSC was established in 1979 under the auspices of the National Institute of Building Sciences to develop and promulgate national regulatory provisions for mitigating earthquake risk to buildings.

Central United States Earthquake Consortium (CUSEC). The CUSEC was established in 1983 to help improve public awareness and education about earthquakes in the central United States; to coordinate multistate planning for preparedness, response, and recovery; and to encourage research in earthquake hazard reduction.

Emergency Management Institute (EMI). This institute of the Federal Emergency Management Agency provides training on mitigation, preparedness, response, recovery, natural and technological hazards, exercise design and evaluation, public information, and other issues through resident and nonresident instruction for federal, state, and local government officials; volunteer organizations; and the private sector.

Humboldt Earthquake Education Center. The Center was established in 1985 as a nonprofit organization within the Department of Geology, Humboldt State University. Its aim is to provide information about seismicity on California's north coast and to foster earthquake awareness, preparedness, and education programs in the schools and among the public.

Institute of Emergency Administration and Planning (IEAP). The IEAP provides teaching, research, and service to the disaster practitioner and research communities, graduating students with B.S. degrees in emergency administration and planning.

Natural Hazards Research and Applications Information Center (NHRAIC). The NHRAIC was established in 1976 at the Institute of Behavioral Science of the University of Colorado, Boulder, as a national clearinghouse for research findings and other information on the social science aspects of natural and related technological hazards.

Southern California and Bay Area Regional Earthquake Preparedness Projects (SCEPP and BAREPP). The SCEPP was created in 1980 to accelerate the pace of earthquake preparedness in Southern California by empowering localities and providing networks of expertise. BAREPP duplicated the project in Northern California and was equally successful. Both projects have been absorbed by the Governor's Office of Emergency Services.

PROFESSIONAL ASSOCIATIONS

Many and diverse professional associations support and foster the engineering, scientific, and practical aspects of hazards mitigation in the United States today. Most of these associations engage in a multitude of activities both to serve their different goals and memberships and to generate an ongoing and growing national commitment to hazards reduction. More often than not, these associations engage in elaborate outreach activities to other groups, organizations, and the public. These are groups that would be partners in future sustainable hazards mitigation networks; their outreach and education capabilities will be invaluable for motivating professionals and the public toward helping create disaster-resilient communities as well as achieving the other goals of sustainability. The few associations listed below illustrate the range of activities that exist today.

American Association of Wind Engineers,
American Geophysical Union,
American Meteorological Society,
American Planning Association,
American Society for Public Administration,
American Society of Civil Engineers,
American Society of Professional Emergency Planners,
Association of American Geographers,
Association of State Dam Safety Officials,
Association of State Wetland Managers,
Association of State Floodplain Managers,
Earthquake Engineering Research Institute,
International Association of Emergency Managers,
National Emergency Management Association,
Research Committee on Disasters,
Seismological Society of America, and
Society for Risk Analysis.

Index

A

B

D

E

F

G

H

I

J

L

M

N

O

P

Q

R

T

U

V

W

Z

BOSTON PUBLIC LIBRARY
3 9999 03695 974 8